The IMA Volumes in Mathematics and its Applications

Volume 62

Series Editors
Avner Friedman   Willard Miller, Jr.

Institute for Mathematics and its Applications
IMA

The **Institute for Mathematics and its Applications** was established by a grant from the National Science Foundation to the University of Minnesota in 1982. The IMA seeks to encourage the development and study of fresh mathematical concepts and questions of concern to the other sciences by bringing together mathematicians and scientists from diverse fields in an atmosphere that will stimulate discussion and collaboration.

The IMA Volumes are intended to involve the broader scientific community in this process.

Avner Friedman, Director
Willard Miller, Jr., Associate Director

* * * * * * * * * *

IMA ANNUAL PROGRAMS

1982–1983	**Statistical and Continuum Approaches to Phase Transition**
1983–1984	**Mathematical Models for the Economics of Decentralized Resource Allocation**
1984–1985	**Continuum Physics and Partial Differential Equations**
1985–1986	**Stochastic Differential Equations and Their Applications**
1986–1987	**Scientific Computation**
1987–1988	**Applied Combinatorics**
1988–1989	**Nonlinear Waves**
1989–1990	**Dynamical Systems and Their Applications**
1990–1991	**Phase Transitions and Free Boundaries**
1991–1992	**Applied Linear Algebra**
1992–1993	**Control Theory and its Applications**
1993–1994	**Emerging Applications of Probability**
1994–1995	**Waves and Scattering**
1995–1996	**Mathematical Methods in Material Science**

IMA SUMMER PROGRAMS

1987	**Robotics**
1988	**Signal Processing**
1989	**Robustness, Diagnostics, Computing and Graphics in Statistics**
1990	**Radar and Sonar (June 18 - June 29)**
	New Directions in Time Series Analysis (July 2 - July 27)
1991	**Semiconductors**
1992	**Environmental Studies: Mathematical, Computational, and Statistical Analysis**
1993	**Modeling, Mesh Generation, and Adaptive Numerical Methods for Partial Differential Equations**
1994	**Molecular Biology**

* * * * * * * * * *

SPRINGER LECTURE NOTES FROM THE IMA:

The Mathematics and Physics of Disordered Media

Editors: Barry Hughes and Barry Ninham
(Lecture Notes in Math., Volume 1035, 1983)

Orienting Polymers

Editor: J.L. Ericksen
(Lecture Notes in Math., Volume 1063, 1984)

New Perspectives in Thermodynamics

Editor: James Serrin
(Springer-Verlag, 1986)

Models of Economic Dynamics

Editor: Hugo Sonnenschein
(Lecture Notes in Econ., Volume 264, 1986)

Paul Van Dooren   Bostwick Wyman
Editors

Linear Algebra for Control Theory

With 2 Illustrations

Springer-Verlag
New York Berlin Heidelberg London Paris
Tokyo Hong Kong Barcelona Budapest

Paul Van Dooren
University of Illinois at
Urbana-Champaign
Coordinated Science Laboratory
1308 W. Main Street
Urbana, IL 61801 USA

Bostwick Wyman
Department of Mathematics
Ohio State University
100 Mathematics Building
231 West 18th Avenue
Columbus, OH 43210-1174 USA

Series Editors:
Avner Friedman
Willard Miller, Jr.
Institute for Mathematics and its
Applications
University of Minnesota
Minneapolis, MN 55455 USA

Mathematics Subject Classifications (1991): 65Fxx, 93Bxx, 15A21, 16D60, 65-06, 93-06

Library of Congress Cataloging-in-Publication Data
Linear algebra for control theory / Paul Van Dooren, Bostwick Wyman, editors.
p. cm. – (The IMA volumes in mathematics and its applications ; v. 62)
Includes bibliographical references.
ISBN 0-387-94267-X (acid-free)
1. Algebras, Linear. 2. Control theory. I. Van Dooren, Paul II. Wyman, B. III. Series.
QA184.L5443 1994 94-249
629.8′312′015125 – dc20

Printed on acid-free paper.

Production managed by Hal Henglein; manufacturing supervised by Jacqui Ashri.
Camera-ready copy prepared by the IMA.
Printed and bound by Braun-Brumfield, Ann Arbor, MI.
Printed in the United States of America.

9 8 7 6 5 4 3 2 1

ISBN 0-387-94267-X Springer-Verlag New York Berlin Heidelberg
ISBN 3-540-94267-X Springer-Verlag Berlin Heidelberg New York

The IMA Volumes
in Mathematics and its Applications

Current Volumes:

Volume 1: Homogenization and Effective Moduli of Materials and Media
Editors: Jerry Ericksen, David Kinderlehrer, Robert Kohn, and J.-L. Lions

Volume 2: Oscillation Theory, Computation, and Methods of Compensated Compactness
Editors: Constantine Dafermos, Jerry Ericksen, David Kinderlehrer, and Marshall Slemrod

Volume 3: Metastability and Incompletely Posed Problems
Editors: Stuart Antman, Jerry Ericksen, David Kinderlehrer, and Ingo Muller

Volume 4: Dynamical Problems in Continuum Physics
Editors: Jerry Bona, Constantine Dafermos, Jerry Ericksen, and David Kinderlehrer

Volume 5: Theory and Applications of Liquid Crystals
Editors: Jerry Ericksen and David Kinderlehrer

Volume 6: Amorphous Polymers and Non-Newtonian Fluids
Editors: Constantine Dafermos, Jerry Ericksen, and David Kinderlehrer

Volume 7: Random Media
Editor: George Papanicolaou

Volume 8: Percolation Theory and Ergodic Theory of Infinite Particle Systems
Editor: Harry Kesten

Volume 9: Hydrodynamic Behavior and Interacting Particle Systems
Editor: George Papanicolaou

Volume 10: Stochastic Differential Systems, Stochastic Control Theory, and Applications
Editors: Wendell Fleming and Pierre-Louis Lions

Volume 11: Numerical Simulation in Oil Recovery
Editor: Mary Fanett Wheeler

Volume 26: Two Phase Flows and Waves
Editors: Daniel D. Joseph and David G. Schaeffer

Volume 27: Nonlinear Evolution Equations that Change Type
Editors: Barbara Lee Keyfitz and Michael Shearer

Volume 28: Computer Aided Proofs in Analysis
Editors: Kenneth Meyer and Dieter Schmidt

Volume 29: Multidimensional Hyperbolic Problems and Computations
Editors Andrew Majda and Jim Glimm

Volume 30: Microlocal Analysis and Nonlinear Waves
Editors: Michael Beals, R. Melrose, and J. Rauch

Volume 31: Mathematics in Industrial Problems, Part 3
by Avner Friedman

Volume 32: Radar and Sonar, Part I
by Richard Blahut, Willard Miller, Jr., and Calvin Wilcox

Volume 33: Directions in Robust Statistics and Diagnostics: Part I
Editors: Werner A. Stahel and Sanford Weisberg

Volume 34: Directions in Robust Statistics and Diagnostics: Part II
Editors: Werner A. Stahel and Sanford Weisberg

Volume 35: Dynamical Issues in Combustion Theory
Editors: P. Fife, A. Liñán, and F.A. Williams

Volume 36: Computing and Graphics in Statistics
Editors: Andreas Buja and Paul Tukey

Volume 37: Patterns and Dynamics in Reactive Media
Editors: Harry Swinney, Gus Aris, and Don Aronson

Volume 38: Mathematics in Industrial Problems, Part 4
by Avner Friedman

Volume 39: Radar and Sonar, Part II
Editors: F. Alberto Grünbaum, Marvin Bernfeld, and
Richard E. Blahut

Volume 40: Nonlinear Phenomena in Atmospheric and Oceanic Sciences
Editors: George F. Carnevale and Raymond T. Pierrehumbert

Volume 41: Chaotic Processes in the Geological Sciences
Editor: David A. Yuen

Volume 42: Partial Differential Equations with Minimal Smoothness and Applications
Editors: B. Dahlberg, E. Fabes, R. Fefferman, D. Jerison, C. Kenig, and J. Pipher

Volume 43: On the Evolution of Phase Boundaries
Editors: Morton E. Gurtin and Geoffrey B. McFadden

Volume 44: Twist Mappings and Their Applications
Editors: Richard McGehee and Kenneth R. Meyer

Volume 45: New Directions in Time Series Analysis, Part I
Editors: David Brillinger, Peter Caines, John Geweke, Emanuel Parzen, Murray Rosenblatt, and Murad S. Taqqu

Volume 46: New Directions in Time Series Analysis, Part II
Editors: David Brillinger, Peter Caines, John Geweke, Emanuel Parzen, Murray Rosenblatt, and Murad S. Taqqu

Volume 47: Degenerate Diffusions
Editors: Wei-Ming Ni, L.A. Peletier, and J.-L. Vazquez

Volume 48: Linear Algebra, Markov Chains, and Queueing Models
Editors: Carl D. Meyer and Robert J. Plemmons

Volume 49: Mathematics in Industrial Problems, Part 5
by Avner Friedman

Volume 50: Combinatorial and Graph-Theoretic Problems in Linear Algebra
Editors: Richard A. Brualdi, Shmuel Friedland, and Victor Klee

Volume 51: Statistical Thermodynamics and Differential Geometry of Microstructured Materials
Editors: H. Ted Davis and Johannes C.C. Nitsche

Volume 52: Shock Induced Transitions and Phase Structures in General Media
Editors: J.E. Dunn, Roger Fosdick, and Marshall Slemrod

Volume 53: Variational and Free Boundary Problems
Editors: Avner Friedman and Joel Spruck

Volume 54: Microstructure and Phase Transitions
Editors: David Kinderlehrer, Richard James, Mitchell Luskin, and Jerry L. Ericksen

Volume 55: Turbulence in Fluid Flows: A Dynamical Systems Approach
Editors: George R. Sell, Ciprian Foias, and Roger Temam

Volume 56: Graph Theory and Sparse Matrix Computation
Editors: Alan George, John R. Gilbert, and Joseph W.H. Liu

Volume 57: Mathematics in Industrial Problems, Part 6
by Avner Friedman

Volume 58: Semiconductors, Part I
Editors: W.M. Coughran, Jr., Julian Cole, Peter Lloyd, and Jacob White

Volume 59: Semiconductors, Part II
Editors: W.M. Coughran, Jr., Julian Cole, Peter Lloyd, and Jacob White

Volume 60: Recent Advances in Iterative Methods
Editors: Gene Golub, Anne Greenbaum, and Mitchell Luskin

Volume 61: Free Boundaries in Viscous Flows
Editors: Robert A. Brown and Stephen H. Davis

Volume 62: Linear Algebra for Control Theory
Editors: Paul Van Dooren and Bostwick Wyman

Forthcoming Volumes:

1991-1992: *Applied Linear Algebra*

Linear Algebra for Signal Processing

1992 Summer Program: *Environmental Studies*

Environmental Studies

1992–1993: *Control Theory*

Robust Control Theory

Control Design for Advanced Engineering Systems:
Complexity, Uncertainty, Information and Organization

Control and Optimal Design of Distributed Parameter Systems

Flow Control

Robotics

Nonsmooth Analysis & Geometric Methods in Deterministic Optimal Control Systems & Control Theory for Power Systems

Adaptive Control, Filtering and Signal Processing

Discrete Event Systems, Manufacturing, Systems, and Communication Networks

Mathematical Finance

1993 Summer Program: *Modeling, Mesh Generation, and Adaptive Numerical Methods for Partial Differential Equations*

1993-1994: *Emerging Applications of Probability*

Probability and Algorithms

Finite Markov Chain Renaissance

Random Discrete Structures

Mathematical Population Genetics

Stochastic Networks

FOREWORD

This IMA Volume in Mathematics and its Applications

LINEAR ALGEBRA FOR CONTROL THEORY

is based on the proceedings of a workshop which was an integral part of the 1991-92 IMA program on "Applied Linear Algebra". We thank Paul Van Dooren and Bostwick Wyman for editing the proceedings. We also take this opportunity to thank the Army Research Office and the National Science Foundation, whose financial support made the workshop possible.

Avner Friedman

Willard Miller, Jr.

PREFACE

During the past decade, the interaction between control theory and linear algebra has been ever increasing, giving rise to new results in both areas. The purpose of this workshop was to develop further and nurture this cross-fertilization by bringing together control theorists and linear algebraists for an exchange of problems, ideas, and techniques. The cross-fertilization between control and linear algebra can be found in the following subfields:

1. Numerical Linear Algebra for Control. One of the most profound impacts of linear algebra in control theory has been in the computational and implementational aspects, where numerical linear algebraic algorithms have strongly influenced the ways in which many problems are being solved. The interaction between the numerical analysis and control community has led to important new results in the solution of special eigenvalue and singular value problems, in calculating invariants and canonical forms, in solving matrix Riccati, Lyapunov, and Sylvester equations and the like. It is still a very active area as new problems keep emerging (see also 3,4, and 5).

2. Canonical Forms and Invariants. Since the introduction of state space models in control theory, canonical forms and invariants have been very important. While there is a clear connection with classical canonical forms such as the Jordan, Kronecker and Smith forms, the problems here are yet more intricate. This is due to the presence of four mappings (often called A,B,C,D) and to a large class of group actions that can be defined on them. Recent interests here include, e.g., the group of orthogonal transforms. Also, for transfer matrix models, new invariants have been defined and relations with other models are being looked at.

3. Ring-theoretic Methods in Linear Control. Many algebraic issues in linear algebra are properly treated through the use of ring and module-theoretic techniques. In the context of control theory, such techniques have already proved extremely useful in the study of various system design questions such as those dealing with invertibility (abstract pole and zero modules), in the study of certain perturbation models (systems with small parameters modeled through matrices over parameter rings), and in questions such as the study of pole shifting for delay and other infinite-dimensional systems through systems over operator rings.

4. Matrix Theory in Control. The models used in control theory are in fact all well-studied objects in matrix theory: rational matrices, polynomial matrices, and pencils. Certain expansions lead to Hankel and Toeplitz matrices, possibly block versions of those. This connection attracted the attention of several researchers in matrix and operator theory, and it resulted in a wealth of new results in matrix theory, which are also of importance to control people. Typical examples are factorizations of rational matrices, embedding and completion problems, invariants, and also canonical forms (see also 2).

5. H^∞ control. Robust control has been one of the most active areas of control during the last five years. The mathematical problems encountered there bear relation to several other fields, but many of the most recent results lie, in fact, in linear algebra. The current trend is to formulate these problems in terms such of state space models, which then leads to special matrix equations. Typical examples are coupled matrix Riccati equations and special singular value problems. Other issues of H^∞ control are also related to the theory of matrix valued functions (see 4), and computational aspects also play an important role here (see 1).

Around each of these subject areas a 1 day session was organized during the workshop. Special thanks ought to go to Angelika Bunse-Gerstner, Volker Mehrmann, Diederich Hinrichsen, Eduardo Sontag, Bostwick Wyman, Leiba Rodman, and, finally, Andre Ran, who skillfully put together nice sessions on each of these topics.

The workshop was a peculiar one in the sense that the five themes looked almost disjoint. The challenge of the workshop was to present the latest results in these areas and to find points of common interest. The results were very positive indeed. Throughout the meeting one could see how the same problems occurred in different sessions and were being tackled from a different point of view. This interaction was also felt strongly in the corridors of the 5th floor of the IMA, where IMA visitors defended their way of looking at a particular common point of interest.

Altogether the main strength of the meeting was its multidisciplinary nature and the cross fertilization resulting from this. The relaxed atmosphere offered by the IMA also played a key role in making this happen. For making this possible, we wish to thank Willard Miller and Avner Friedman from the IMA, and also Dick Brualdi, who was a key figure in the organization of the Applied Linear Algebra Year held at the IMA.

Paul Van Dooren
Bostwick Wyman

CONTENTS

RECURSIVE MODELING OF DISCRETE-TIME TIME SERIES*

A.C. ANTOULAS[†]

Abstract. The approach to modeling inspired by the behavioral framework consists in treating all measurements on an equal footing, not distinguishing between inputs and outputs. Consequently, the initial search is for autonomous models. In the linear, time invariant case, the main result guarantees the existence of a *minimal complexity autonomous generating model* Θ^*. This means that all other models can be explicitely constructed from Θ^*. Among them in most cases, the so-called *input-output controllable models* are of interest. The main purpose of this paper is to show how these models can be constructed in a an easy-to-implement, recursive way.

1. Introduction. We are given several discrete-time vector-valued time series

$$\mathbf{w}_i, i \in \underline{\mu}.$$

The problem is to determine linear, time invariant *exact* models compatible with (i.e. which could have generated) these measurements. An important special case occurs when the $\mathbf{w}_i$ consist of inputs $\mathbf{u}_i$ and outputs $\mathbf{y}_i$, i.e.

$$\mathbf{w}_i = \begin{pmatrix} \mathbf{u}_i \\ \mathbf{y}_i \end{pmatrix}, \; i \in \underline{\mu}.$$

Of primary interest in this case, are the so-called *input-output (I/O)* models, which explain how the output $\mathbf{y}_i$ depends on the input $\mathbf{u}_i$.

The approach we will follow is in the spirit of the behavioral framework as developed by Willems, for example, in [7]. The theoretical aspects are de-emphasized in favor of the algorithmic ones. Thus initially, we will *not* distinguish between inputs and outputs, even if these are explicitely given. Instead we will treat the $\mathbf{w}_i$ as measurements coming from an autonomous system and will seek to determine **autonomous models**. It turns out that there exists a *unique* (up to equivalence) autonomous model Θ^* of minimal complexity generating the measurements $\mathbf{w}_i$. The main result is that *all* models Θ of these measurements, to wit

- autonomous models (i.e. models explaining $\mathbf{w}_i$),
- input models (i.e. models explaining $\mathbf{u}_i$),
- output models (i.e. models explaining $\mathbf{y}_i$),
- input-output models (i.e. models explaining $\mathbf{y}_i$ in terms of $\mathbf{u}_i$),

satisfying properties like

- controllability and
- causality,

---

* Partially supported by NSF through Grant ECS-9017618 and by the State of Texas through the Advanced Technology Program Grant 003604-041.

† Department of Electrical and Computer Engineering, Rice University, Houston, Texas 77251-1892, USA. e-mail: aca@rice.edu - fax: +1-713-524-5237.

can be readily obtained from the unique autonomous model Θ^* mentioned above.

A general framework for treating exact modeling problems was set up in Antoulas and Willems [3]. The present paper is an application of this framework to the discrete-time modeling problem. The treatment however is self-contained and does not require knowledge of [3] to follow. There are two important special cases of the discrete time series modeling problem which merit to be mentioned. The first is the *realization problem.* Here the data set consists of the response to impulses (one applied to each input channel). The complete recursive solution of this problem was given in Antoulas [1]. The approach presented below uses a slightly modified main tool (the generating system Θ^* instead of the $\mathbf{W}$-matrix). This provides the added flexibility of being able to treat tangential (directional) versions of the realization problem (cf. example **B**, at the end of the paper); a study of state space aspects of this latter problem can be found in Gohberg, Kaashoek, and Lerer [5]. A second special case is that of the multiple-point tangential interpolation problem, at the origin. Thus, the results of Antoulas et al. [2] are closely related to the ones given below.

The paper is structured as follows. The data set $\mathbf{D}$ and the model set $\mathcal{M}(\mathbf{D})$ are defined in sections 2 and 3, respectively. Section 4 deals with the definition of various sub-families of models and of model complexity. The main result theorem 5.1, is presented in the next section. It is an existence result stating that the class $\mathcal{M}(\mathbf{D})$ is generated by a single element Θ^*, which happens to be an autonomous system; thus every model Θ can be obtained as a linear combination of the rows of this fixed autonomous model Θ^*. It is also argued that the class of controllable models, denoted by $\mathcal{M}_{contr}(\mathbf{D})$ is much more intricate to characterize. However, given Θ^* and one controllable model Θ_{contr} of a given (say, minimal) complexity, all other models of that same complexity can be characterized in a straightforward way.

Section 6 shows that the invariant factors of Θ^*, are powers of s, the value of these powers being directly related to the length of each time series in $\mathbf{D}$. The first part of section 7 is devoted to the recursive construction of Θ^*, while the second part is concerned with the simultaneous update of both Θ^* and of a minimal complexity controllable model Θ^*_{contr}. A consequence of the theory is that the controllable models *cannot* be updated without updating the autonomous generating system Θ^*. Thus, the **generating system** turns out to be the **natural tool** for dealing with the modeling problem at hand.

2. The data. We will consider discrete-time vector-valued time series

$$\mathbf{v}_i : \mathbf{Z} \rightarrow \mathbf{R}^q, \quad i \in \underline{\mu}, \quad q \geq 1,$$

which are at rest in the (distant) past:

$$\mathbf{v}_i(t) = 0, \quad t < t_i, \quad \mathbf{v}_i(t_i) \neq 0, \quad i \in \underline{\mu}.$$

This last assumption can be made without loss of generality (see [3], section 2.E). The problem we will study is to model these time series up to some arbitrary finite time $T_i \geq t_i$, $i \in \underline{\mu}$, by means of *linear* and *time invariant* systems. Because of time invariance, the models will not explicitely depend on t_i or T_i, but only on their difference

$$\kappa_i := T_i - t_i, \; i \in \underline{\mu}.$$

In order to have the time series defined on the same domain, they will be shifted so that T_i is at the origin:

$$\mathbf{w}_i := \sigma^{-T_i}\mathbf{v}_i, \quad i \in \underline{\mu},$$

where σ^{-1} denotes the *forward shift*:

$$\sigma^{-1}f(t) := f(t-1), \quad t \in \mathbf{Z}.$$

The *data set* is thus made up of time series defined on the negative integers $\mathbf{Z}_-$:

$$\mathbf{D} \; := \; \{\mathbf{w}_i \, : \, \mathbf{Z}_- \to \mathbf{R}^q; \; \mathbf{w}_i(-\kappa_i) \neq 0; \; \mathbf{w}_i(t) = 0, \; t < -\kappa_i; \; i \in \underline{\mu}\}. \tag{2.1}$$

The length of $\mathbf{w}_i$ is κ_i+1, and the sum

$$\nu \; := \; \sum_{i \in \underline{\mu}} (\kappa_i + 1). \tag{2.2}$$

is the compound length of the time series in $\mathbf{D}$. In the sequel we will also make use of the *shifted data set*, which is the closure of the data set under the operation of forward shifting:

$$\mathbf{D}_\sigma \; := \; \{\mathbf{w}_j^i := \sigma^{-i}\mathbf{w}_j \, : \, \mathbf{Z}_- \to \mathbf{R}^q; \; i \geq 0, \; j \in \underline{\mu}\}. \tag{2.3}$$

3. The model class. The power set Y^X denotes the set of all maps with domain X and range Y. The models we will be considering are operators

$$\Theta : \; (\mathbf{R}^q)^{\mathbf{Z}_-} \to (\mathbf{R}^q)^{\mathbf{Z}_-},$$

which are *linear* and *time invariant* and can be represented as a finite polynomial in the forward shift

$$\Theta(\sigma^{-1}) \; := \; \Theta_0 \; + \; \Theta_1\sigma^{-1} \; + \; \cdots \; + \; \Theta_L\sigma^{-L};$$

the coefficients are constant matrices with q columns: $\Theta_i \in \mathbf{R}^{\cdot \times q}$. We will thus use the notation

$$\Theta \in \mathbf{R}^{\cdot \times q}[s],$$

where the indeterminate s is to be interpreted as σ^{-1}. To each Θ we will also associate its *behavior*, which is a set of time series:

$$\mathbf{B}(\Theta) \ := \ ker\ \Theta \ = \ \{\mathbf{w} \in (\mathbf{R}^q)^{\mathbf{Z}_-} : \ \Theta(\sigma^{-1})\mathbf{w} \ = \ 0\}. \tag{3.1}$$

Clearly, $\mathbf{B}(\Theta)$ is linear, and since

$$\sigma^{-T}\mathbf{B}(\Theta) \subset \mathbf{B}(\Theta) \ \ T > 0,$$

it is also shift invariant. The behavior of Θ can thus be given the structure of an $\mathbf{R}[s]$-module, where s is the forward shift. The *family of models* of the data set $\mathbf{D}$ consists of all Θ such that $\Theta(\sigma^{-1})\mathbf{w}_i = 0$, $i \in \underline{\mu}$:

$$\mathcal{M}(\mathbf{D}) \ := \ \{\Theta \in \mathbf{R}^{\cdot \times q}[s] : \ \mathbf{B}(\Theta) \supset \mathbf{D}\} \ \subset \ \mathbf{R}^{\cdot \times q}[s]. \tag{3.2}$$

Thus, if the behavior includes tha data set, it includes the shifted data set as well:

$$\mathbf{B}(\Theta) \supset \mathbf{D} \ \Rightarrow \ \mathbf{B}(\Theta) \supset \mathbf{D}_\sigma. \tag{3.3}$$

The following standard results will be used in the sequel. For a proof we refer to Antoulas and Willems [3], proposition 3.10.

PROPOSITION 3.1.
Let $\mathbf{B}(\Theta) := ker\,\Theta$, where $\Theta \in \mathbf{R}^{\cdot \times q}[s]$.

(a) $\mathbf{B}(\Theta_1) \subseteq \mathbf{B}(\Theta_2)$ if, and only if, $\Theta_2 = \Gamma\Theta_1$, for some polynomial matrix Γ of appropriate dimension. If Γ is unimodular, equality holds. Conversely, if Θ_1 and Θ_2 have full row rank, then Γ is unimodular. Finally, assuming that Θ has full row rank, $\mathbf{B}(\Theta) = 0$ if, and only if, Θ is unimodular.

(b) $\mathbf{B}(\Theta_1) \cap \mathbf{B}(\Theta_2) = \mathbf{B}(\Theta_{12})$, where Θ_{12} is the greatest common right divisor of Θ_1 and Θ_2:

$$\Theta_{12} \ = \ gcrd\ (\Theta_1,\ \Theta_2).$$

(c) $\mathbf{B}(\Theta_1) + \mathbf{B}(\Theta_2) \ = \ \mathbf{B}(\Theta_{12})$, where Θ_{12} is the least common left multiple of Θ_1 and Θ_2:

$$\Theta_{12} \ = \ lclm\ (\Theta_1,\ \Theta_2).$$

4. Classification of models. Among all models we will single out three important classes: that of *autonomous*, that of *controllable* and that of *input-output* models. The importance of the former derives from the fact that the family $\mathcal{M}(\mathbf{D})$ is generated by a single autonomous model of minimal complexity. This pivotal result is presented in theorem 5.1. The importance of the other two properties stems from the fact that they lead to the usual class of models; input-output controllable models are, namely, those which can equivalently be described in terms of *transfer functions*.

4.1. Autonomous, I/O, and controllable models. Let the external variables be partitioned, possibly after permutation, into two groups:

$$\mathbf{w} = \begin{pmatrix} \mathbf{u} \\ \mathbf{y} \end{pmatrix}, \quad \mathbf{u} \in (\mathbf{R}^m)^{\mathbf{Z}_-}, \quad \mathbf{y} \in (\mathbf{R}^p)^{\mathbf{Z}_-}, \quad m + p = q.$$

Let the model Θ be partitioned accordingly

$$\Theta \;=\; (Q \quad -T) \in \mathbf{R}^{g\times q}[s], \quad Q \in \mathbf{R}^{g\times m}[s], \quad T \in \mathbf{R}^{g\times p}[s]. \tag{4.1}$$

The expression $\Theta(\sigma^{-1})\mathbf{w} = 0$, can in this case be rewritten as

$$T(\sigma^{-1})\mathbf{y} \;=\; Q(\sigma^{-1})\mathbf{u}.$$

If the following conditions hold

$$g \;=\; p \quad \text{and} \quad det\ T \neq 0, \tag{4.2}$$

for every $\mathbf{u} \in (\mathbf{R}^m)^{\mathbf{Z}_-}$, there exists $\mathbf{y} \in (\mathbf{R}^p)^{\mathbf{Z}_-}$, such that the above relationship is satisfied. In this case, $\mathbf{u}$ is *free* or an *input*, and $\mathbf{y}$ is *bound* or an *output*. Such Θ will be called *input-output* (*I/O*) models. If, however

$$g \;=\; q \quad \text{and} \quad det\ \Theta \neq 0,$$

all components of $\mathbf{w}$ are bound and none can be chosen freely. Such models will be called *autonomous*. Having declared $\mathbf{u}$ as an input and $\mathbf{y}$ as an output, if

$$g \;=\; m, \quad T \;=\; 0 \quad \text{and} \quad det\ Q \neq 0,$$

we are dealing with an *input* model, while if

$$g \;=\; p, \quad Q \;=\; 0 \quad \text{and} \quad det\ T \neq 0,$$

we have an *output* model.

An important property of models is that of *controllability*. $\Theta \in \mathbf{R}^{p\times q}[s]$, is a *controllable* system iff all of its $\hat{p}$ non-zero invariant factors are equal to one, i.e.,

$$rank\ \Theta(\lambda) \;=\; \hat{p} \leq p, \quad \forall \lambda \in \mathbf{C}. \tag{4.3}$$

This condition is a generalization of the Hautus controllability test, to the case of systems where inputs and outputs may not yet have been defined. It follows that $\Theta = (Q \quad -T) \in \mathbf{R}^{p\times q}[s]$, is a *controllable input-output* system iff $det\ T \neq 0$ and $rank\ \Theta(\lambda) = p$, $\forall \lambda \in \mathbf{C}$. This is equivalent to the well known left coprimeness of T, Q (see Fuhrmann [4]):

$$gcld\ (T,\ Q) = I_q,$$

(*gcld* stands for "greatest common left divisor"). Controllable input-output systems can be equivalently described in terms of their *transfer function*: $Z := T^{-1}Q$, which is a rational function in s. Finally, an input-output system is *causal*, i.e. $\mathbf{y}(t)$ depends on $\mathbf{u}(\tau)$, for $\tau \leq t$, iff Z is proper rational in $\sigma = s^{-1}$, irrespective of whether the system is controllable or not.

For the modeling problem at hand, as a consequence of theorem 5.1, the properties of controllability (4.3) and causality, become

$$rank\ \Theta(0) = \hat{p} \quad \text{and} \quad det\ T(0) \neq 0, \tag{4.4}$$

respectively. The following lemma is again a consequence of theorem 5.1.

LEMMA 4.1. *Given* $\mathbf{D}$, *and an arbitrary partitioning of* $\mathbf{w}$, *there always exist models* $\Theta \in \mathcal{M}(\mathbf{D})$, $\Theta \in \mathbf{R}^{g\times q}[s]$, *which are* **(i)** *minimal complexity input-output, for all* $g < q$; **(ii)** *minimal complexity controllable, for all* $g \leq p$, *where* $m = q - p$ *is the number of linearly independent time series, and* p *is the number of unity invariant factors of* Θ^*. **(iii)** *There exist causal models iff* $rank\,[\mathbf{u}_1(-\kappa_1)\ \ \ldots\ \ \mathbf{u}_\mu(-\kappa_\mu)] = \mu$. *Indeed, if this condition is satisfied, all controllable input-output models are causal.*

REMARK 4.1. Autonomous and controllable models can equivalently be described by means of their behaviors. An autonomous behavior is characterized by the fact that its trajectories are completely determined by their past (with respect to an arbitrary time t_0). Controllable behaviors occupy the opposite end of the spectrum: every past can be concatenated (at some t_0) with any (desired) future to yield a trajectory belonging to the same behavior. This aspect of the theory will not be pursued here. We refer to Antoulas and Willems [3] for details. □

4.2. Model complexity. Recall definition (3.1) of the behavior of systems in $\mathbf{R}^{\cdot\times q}[s]$. Two such systems are *equivalent* iff they have the same *behavior*:

$$\Theta_1 \sim \Theta_2 \ \Leftrightarrow\ \mathbf{B}(\Theta_1) = \mathbf{B}(\Theta_2). \tag{4.5}$$

If Θ_1 and Θ_2 have full row rank, according to proposition 3.1(a), they are equivalent iff they are equal up to left multiplication by a unimodular matrix.

Thus every model $\hat{\Theta} \in \mathbf{R}^{g\times q}[s]$, is equivalent to a model $\Theta \in \mathbf{R}^{p\times q}[s]$, having full row rank p over $\mathbf{R}[s]$. Furthermore, Θ can be assumed to be *row reduced*[1] with row indices ν_i, $i \in \underline{p}$. The *complexity* $\delta(\Theta)$ is now defined as a pair of non-negative integers:

$$\delta(\Theta) := (m,\ n), \quad m := q - p, \quad n := \sum_{i\in \underline{p}} \nu_i. \tag{4.6}$$

p is equal to the number of linearly independent equations defining the system, and in the context of input-output models, m is the least number of inputs, i.e. the least number of *unexplained* variables, of the model. The second integer n, gives the number of states required in a (minimal) state-variable representation.

In order to compare systems, we will make use of the *lexicographic ordering*:

$$(m_1, n_1) \leq (m_2, n_2) \quad \Leftrightarrow \quad \text{either } m_1 < m_2 \text{ or } m_1 = m_2 \text{ and } n_1 \leq n_2,$$

which provides a total ordering in $\mathbf{R}^{\cdot\times q}[s]$. This ordering is but one way of comparing the complexity of models. By minimizing it, one minimizes first the number of *unexplained* or *free* variables, and then the number of state variables. Thus *any* autonomous model (model with no inputs $\Rightarrow\ m = 0$) has lower complexity than *any* non-autonomous model. It will be shown later (cf. remark 6.1) that the least number of unexplained variables in controllable models is equal to the number of linearly independent measurements in $\mathbf{D}$.

5. Characterization of models.

In this section we will present the main result. It states that the model set $\mathcal{M}(\mathbf{D})$ is generated by an autonomous system Θ^*, which is unique up to equivalence, and will be referred to as a *generating system*. The issue of characterizing the sub-class of controllable models denoted by $\mathcal{M}_{contr}(\mathbf{D})$ is also addressed.

THEOREM 5.1. *Consider the data set* $\mathbf{D}$ *defined by (2.1) and the associated model set* $\mathcal{M}(\mathbf{D})$ *defined by (3.2).*

(a) The greatest common right divisor Θ_{12} *of any two models* Θ_1, Θ_2, *is a model:*

$$\Theta_1,\ \Theta_2 \in \mathcal{M}(\mathbf{D}) \quad \Rightarrow \quad \Theta_{12} \in \mathcal{M}(\mathbf{D}).$$

(b) The model set is generated by an **autonomous** *model* Θ^*, *which is unique modulo equivalence and its determinant is a power of* s*:*

[1] For a definition of row reducedness see, e.g., Kailath [6]

$$\mathcal{M}(\mathbf{D}) := \mathbf{R}^{\cdot \times q}[s] \cdot \Theta^* := \{\Gamma\Theta^* : \Gamma \in \mathbf{R}^{\cdot \times q}[s],\ det\, \Theta^* = s^\kappa,\ \kappa \le \nu\}, \tag{5.1}$$

where ν is defined by (2.2). Furthermore, among all models, the generating system Θ^* has minimal complexity in the lexicographic ordering

$$\delta(\Theta^*) = (0,\ \kappa) \le \delta(\Theta), \quad \forall\, \Theta \in \mathcal{M}(\mathbf{D}). \tag{5.2}$$

REMARK 5.1. The behavioral aspects of the above result are as follows. Let $\mathbf{B}^*$ be the linear space generated by the elements of the shifted data set

$$\mathbf{B}^* := span_{\mathbf{R}}\, \mathbf{D}_\sigma. \tag{5.3}$$

Clearly, $\mathbf{B}^*$ is also shift invariant. Recall definition (3.1). From proposition 3.1, part *(b)*, and the above theorem, it follows that $\mathbf{B}^*$ is equal to $\mathbf{B}(\Theta^*)$, and is the smallest linear, shift-invariant cover of $\mathbf{D}$; furthermore, it is contained in the behavior of every other model:

$$\mathbf{D} \subset \mathbf{B}^* = \mathbf{B}(\Theta^*) \subset \mathbf{B}(\Theta), \quad \forall\, \Theta \in \mathcal{M}(\mathbf{D}). \tag{5.4}$$

For details on these aspects of the behavioral approach to exact modeling, see Antoulas and Willems [3]. □

Proof. Since we are dealing with linear and shift-invariant models, $\Theta_1,\ \Theta_2 \in \mathcal{M}(\mathbf{D})$, implies that $A\Theta_1 + B\Theta_2 \in \mathcal{M}(\mathbf{D})$, for all polynomial matrices A, B of appropriate size. Hence, the greatest common right divisor of these two models will also be a model. This proves *(a)*. It follows that the greatest common right divisor of all models denoted by Θ^*, is a model which is unique up to left multiplication by a unimodular matrix. Finally, recalling (2.1), $s^{\kappa^*} I_q$ is clearly a model, where $\kappa^* := max\,\{\kappa_i, i \in \underline{\mu}\}$; this implies that $\kappa \le q\kappa^*$. It will follow from theorem 6.1, that $deg\, det\, \Theta^* \le \nu$. Hence, Θ^* is square, non-singular, and its determinant is a power of s, bounded by $\nu \le q\kappa^*$. □

In contrast to $\mathcal{M}(\mathbf{D})$, the set of *controllable* models denoted by $\mathcal{M}_{contr}(\mathbf{D}) \subset \mathcal{M}(\mathbf{D})$, is *not* generated by a single model. The reason is that the greatest common divisor of controllable models need not be controllable. This is illustrated next.

Example. Let $\mathbf{D} = \{\mathbf{w}\}$ where

$$\mathbf{w}(t) := 0,\ t < -1,\ \mathbf{w}(-1) := \begin{pmatrix} 1 \\ -1 \end{pmatrix},\ \mathbf{w}(0) := \begin{pmatrix} 0 \\ 1 \end{pmatrix};$$

clearly, $\Theta_1 := (1 \quad 1+s)$, $\Theta_2 := (1-s \quad 1) \in \mathcal{M}(\mathbf{D})$. By (4.4) these models are controllable. However, their greatest common right divisor is

$$gcrd(\Theta_1,\, \Theta_2) \;=\; \begin{pmatrix} 1 & 1+s \\ 1-s & 1 \end{pmatrix} \;=\; \Theta^*,$$

has complexity $(0,2)$, and is autonomous. A simple argument shows that Θ^* is a generating system for $\mathcal{M}(\mathbf{D})$. The family of controllable models of minimal complexity $(1,1)$ can be parametrized as an arbitrary linear combination of the rows of Θ^*, in terms of the real parameters α, β:

$$\Theta^*_{contr}(s;\alpha,\beta) \;:=\; [\alpha+\beta(1-s) \qquad \alpha(1+s)+\beta], \quad \alpha+\beta\neq 0.$$

Actually, the family of all controllable models of complexity $(1,n)$, $n>1$, can be parametrized by letting α and β in the above expression, be arbitrary polynomials of degree $n-1$, satsfying $\alpha(0)+\beta(0)\neq 0$. Clearly, none of the models with complexity $(1,n)$ is generated by a model with complexity $(1,\hat{n})$, $\hat{n}<n$. □

In general the following result holds. Let (m,n) be the minimal complexity of controllable models. We will denote by $\mathcal{M}_{m,n+i}$ the set of all controllable models of complexity $(m,n+i)$. The *generating set* for $\mathcal{M}_{contr}(\mathbf{D})$ is

$$\mathcal{M}^*_{contr}(\mathbf{D}) \;:=\; \bigcup_{i\geq 0} \mathcal{M}_{m,n+i}$$

Notice that each $\mathcal{M}_{m,n+i}$ contains models which are *not* equivalent. Thus, every model in $\mathcal{M}_{contr}(\mathbf{D})$ can be obtained as a linear combination of the rows of some model belonging to the above generating set. Despite the complex nature of $\mathcal{M}^*_{contr}(\mathbf{D})$, the important fact to keep in mind is that, as illustrated in the example, *given Θ^* and one controllabe model Θ_{contr} of given complexity, all other models of the same complexity can be explicitely parametrized.* For more details we refer to Antoulas and Willems [3], section 9.1.

6. The invariant factors of Θ^*. Before discussing the recursive construction of generating systems, we will compute their invariant factors (see theorem 6.1). In the process, a *non-recursive* way of constructing generating systems will emerge, shedding light into their structure. To simplify the exposition, we will assume without loss of generality, that the measurements are *linearly independent.* This means that the non-zero entries of the shifted data set $\mathbf{D}_\sigma$ form a basis for $\mathbf{B}^*$ defined by (5.3). This is equivalent to the linear independence of the time series $\sigma^{-\kappa_i}\mathbf{w}_i$, $i\in\underline{\mu}$:

$$rank\,[\mathbf{w}_1(-\kappa_1) \;\;\ldots\;\; \mathbf{w}_\mu(-\kappa_\mu)] \;=\; \mu\leq q. \tag{6.1}$$

If this condition is not satisfied, one has to search for a basis of $\mathbf{B}^*$. We refer to Antoulas and Willems [3], section 4, for an explicit way of doing this.

Case $q = 1$. Let $\mathbf{D} = \{\mathbf{w}\}$, be given, where

$$\mathbf{w} := [\cdots 0 \cdots 0 \;\; \mathbf{w}(-\kappa) \;\; \mathbf{w}(-\kappa+1) \cdots \mathbf{w}(-1) \;\; \mathbf{w}(0)], \quad \mathbf{w}(-\kappa) \neq 0.$$

Clearly, $\sigma^{-i}\mathbf{w} = 0$, for all $i > \kappa$, and a generating system is the monic polynomial

$$\Theta^* = s^{\kappa+1}.$$

Furthermore according to (5.1), the set of all models of $\mathbf{w}$ consists of all multiples of $s^{\kappa+1}$.

Case $q > 1$. By theorem 5.1(b), the determinant of Θ^* is a power of s. Consequently, the invariant factors of Θ^* are powers of s. The following result shows that if (6.1) is satisfied these powers are exactly the κ_i, defined by (2.1).

THEOREM 6.1. *Given the data set $\mathbf{D}$ defined by (2.1), and satisfying the linear independence condition (6.1), the invariant factors of any generator Θ^* of $\mathcal{M}(\mathbf{D})$ are:*

$$\boxed{s^{\kappa_1+1}, \; s^{\kappa_2+1}, \; \cdots, \; s^{\kappa_\mu+1}, \; \underbrace{1, \; \cdots, \; 1}_{(q-\mu)\ terms}.}$$

In general, $\kappa_i + 1$, $i \in \underline{\mu}$, constitute upper bounds for the exponents of the invariant factors of the generating systems.

From the above analysis, the proof of this theorem is clear for $q = 1$. In the general case the proof is based on a reduction to the case $q = 1$. To that effect, the following auxiliary result will be used. It reduces the time series $\mathbf{w}_i$, $i \in \underline{\mu}$, to a kind of simultaneous upper triangular form.

PROPOSITION 6.2. *Consider $\mathbf{D}$ defined by (2.1), and assume without loss of generality, that $\kappa_i \leq \kappa_{i+1}$, $i \in \underline{\mu-1}$. There exists a unimodular $U \in \mathbf{R}^{q\times q}[s]$, such that $\hat{\mathbf{D}} := U\mathbf{D}$, is made up of the time series $\hat{\mathbf{w}}_i := U(\sigma^{-1})\mathbf{w}_i$, $i \in \underline{\mu}$, which satisfy*

$$\hat{\mathbf{w}}_i(-\kappa_i) = e_i, \; \hat{\mathbf{w}}_{j,i}(t) = 0 \;\; for \;\; j > i, \; \forall t \in \mathbf{Z}_- \;\; and \;\; \forall j \in \underline{q}, \; t < -\kappa_i,$$

where e_i is the ith unit vector, and $\hat{\mathbf{w}}_{j,i}$ denotes the jth component of the ith time series.

COROLLARY 6.3. *From the triangular form of the new data set $\hat{\mathbf{D}}$, it readily follows that*

$$\Delta^* := diag\,(s^{\kappa_1+1}, \; s^{\kappa_2+1}, \; \cdots, \; s^{\kappa_\mu+1}, \; \underbrace{1, \; \cdots, \; 1}_{(q-\mu)\ terms}). \tag{6.2}$$

is a generating system for the model set $\mathcal{M}(\hat{\mathbf{D}})$. *Consequently,* $\Delta^* U$ *is a generating system for the original data set* $\mathbf{D}$.

REMARK 6.1. Since

$$q-\mu \;=\; rank\ (\Delta^* U)(0) \leq rank\ (\Delta^* U)(\lambda),\ \forall \lambda \in \mathbf{C},$$

it follows from the definition of controllabillity (4.3), that controllable models contain at most $q-\mu$ equations, i.e. in an input-output model at most $q-\mu$ variables can be explained as functions of the inputs. Therefore, the remaining μ variables will be *unexplained*. In terms of the quantities defined above, a parametrization of controllable models is given by

$$\Theta^*_{contr} \;=\; [A \quad B]\begin{bmatrix} \hat{\Delta} & 0 \\ 0 & I_{q-\mu} \end{bmatrix} U \in \mathbf{R}^{(q-\mu)\times q}[s],$$

where $\hat{\Delta}$ is the diagonal matrix composed of the non-unity invariant factors of Δ^*, $det\ B(0) \neq 0$, and A, B are left coprime polynomial matrices. □

Proof of corollary 6.3 and theorem 6.1. Since $\hat{\mathbf{D}} = U\mathbf{D}$, any model of $\hat{\mathbf{D}}$ is also a model for $\mathbf{D}$. Moreover, because of the unimodularity of U, the fact that Δ^* is a generating system of $\hat{\mathbf{D}}$, implies that $\Delta^* U$ is a generating system for $\mathbf{D}$. □

Proof of proposition 6.2. The first step consists in reducing the time series $\mathbf{w}_1(t), -\kappa_1 \leq t \leq 0$, into multiples of the first unit vector e_1. This will be done iteratively. Since $\mathbf{w}_1(-\kappa_1) \neq 0$, there exists a constant nonsingular matrix U_0 of size q, such that

$$U_0 \mathbf{w}_1(-\kappa_1) = e_1.$$

Let

$$U_0 \mathbf{w}_1(-\kappa_1+1) \;=:\; \begin{pmatrix} x_1 \\ y_1 \end{pmatrix},\quad x_1 \in \mathbf{R},\quad y_1 \in \mathbf{R}^{q-1}.$$

Define the unimodular matrix

$$U_1(s) \;:=\; \begin{pmatrix} 1 & 0 \\ -y_1 s & I_{q-1} \end{pmatrix} \in \mathbf{R}^{q\times q}[s].$$

It follows that

$$U_1(\sigma^{-1}) U_0 \left[\mathbf{w}_1(-\kappa_1) \quad \mathbf{w}_1(-\kappa_1+1)\right] \;=\; e_1[1 \quad x_1].$$

Assuming that

$$U_1(\sigma^{-1}) U_0 \mathbf{w}_1(-\kappa_1+2) \;=:\; \begin{pmatrix} x_2 \\ y_2 \end{pmatrix},\quad x_2 \in \mathbf{R},\quad y_2 \in \mathbf{R}^{q-1},$$

and defining the unimodular matrix

$$U_2(s) := \begin{pmatrix} 1 & 0 \\ -y_2 s^2 & I_{q-1} \end{pmatrix} \in \mathbf{R}^{q\times q}[s],$$

the following relationship holds true

$$U_2(\sigma^{-1})U_1(\sigma^{-1})U_0\,[\mathbf{w}_1(-\kappa_1)\ \ \mathbf{w}_1(-\kappa_1+1)\ \ \mathbf{w}_1(-\kappa_1+2)] = e_1[1\ \ x_1\ \ x_2].$$

Proceeding in a similar way, a unimodular matrix

$$U^{(1)} := U_{\kappa_1} \cdots U_2 U_1 U_0$$

can thus be constructed so that only the first component of $\hat{\mathbf{w}}_1$ is non-zero for $-\kappa_1 \leq t \leq 0$. Disregarding the first component of all the time series, we are left with $\mu-1$ non-zero time series having $q-1$ components. We repeat the above procedure in order to reduce the first time series of this new set to the required form, by means of the unimodular matrix $\overline{U}^{(2)}$. Continuing this procedure the required U is the product

$$U := U^{(\mu)} \cdots U^{(2)} U^{(1)},$$

where $U^{(i)} := diag\,(I_{i-1},\ \overline{U}^{(i)}),\ i = 2, \cdots, \mu$. □

7. Recursive update of the model set $\mathcal{M}(\mathbf{D})$. We are now ready to show how the model set $\mathcal{M}(\mathbf{D})$ can be constructed. The construction will be *recursive* with respect to the number of measurements of the jth time series, where the jth can be one which is currently being modeled, or some other time series. By theorem 5.1 we know that all models in $\mathcal{M}(\mathbf{D})$ can be obtained as linear combinations of the rows of a generating system. The first sub-section below is devoted to the recursive update of a generating system. Following the remarks at the end of section 5, the construction of the class of controllable models of given complexity, requires a generating system and one controllable model of the desired complexity. The second sub-section describes the simultaneous update of a generating system and of one controllable model of minimal complexity.

7.1. Recursive update of a generating model. By time invariance, if the time series $\mathbf{w}$ has been modeled by some system Θ, *all* of its shifted versions $\sigma^{-i}\mathbf{w}$, $i \geq 0$, have been modeled as well (cf. (3.3)). Therefore, recursive modeling of the data set $\mathbf{D}$, amounts to the succesive modeling of subsets Ψ of the shifted data set $\mathbf{D}_\sigma$, which are *shift invariant*. Thus let Ψ and $\hat{\Psi}$ be *shift invariant* subsets of $\mathbf{D}_\sigma$, satisfying

$$\hat{\Psi} := \Psi \cup \{\mathbf{w}_j^i\},\ \mathbf{w}_j^i \in \mathbf{D}_\sigma. \tag{7.1}$$

Our goal in this section is to show how a generating system Θ^* of Ψ, can be updated to account for the new time series $\mathbf{w}_j^i$, thus providing a generating system $\hat{\Theta}^*$ for $\hat{\Psi}$. Towards this goal, we define the error time series

$$\text{(7.2)} \qquad \mathbf{e} := \Theta^*(\sigma^{-1})\mathbf{w}_j^i.$$

Because the set Ψ is shift invariant, only the value of $\mathbf{e}$ at time $t = 0$ might be different from zero. With $\Theta^*(s) := \Theta_0 + \Theta_1 s + \Theta_2 s^2 + \cdots$, we define

$$\text{(7.3)} \qquad \epsilon := \mathbf{e}(0) = \Theta_0 \mathbf{w}_j^i(0) + \Theta_1 \mathbf{w}_j^i(-1) + \Theta_2 \mathbf{w}_j^i(-2) + \cdots .$$

If $\epsilon = 0$, $\Gamma := I_q$, is a generating system for the error time series $\mathbf{e}$. Otherwise, let the first non-zero entry of ϵ from the top, be the rth:

$$\text{(7.4)} \qquad \epsilon := \begin{pmatrix} 0 \\ 1 \\ \hat{\epsilon} \end{pmatrix} \in \mathbf{R}^q, \quad \hat{\epsilon} \in \mathbf{R}^{q-r}.$$

In this case a generating system is given by:

$$\text{(7.5)} \qquad \Gamma(s) := \begin{pmatrix} I_{r-1} & & \\ & s & \\ & -\hat{\epsilon} & I_{q-r} \end{pmatrix} \in \mathbf{R}^{q \times q}[s],$$

where all the missing entries are zeros. The following result provides the desired *update* of the generating system Θ^*.

THEOREM 7.1. *With the notation introduced above, if* Θ^* *is a generating system for* Ψ

$$\boxed{\hat{\Theta}^* := \Gamma \Theta^*}$$

is a **generating system** *for* $\hat{\Psi}$.

Proof. If $\epsilon = 0$, an update is not necessary, and $\hat{\Theta}^* = \Theta^*$. This can only occur if $i = \kappa_j$, *and* $\mathbf{w}_j^i$ is linearly dependent on the time series belonging to Ψ, i.e., (6.1) does not hold for the time series belonging to $\hat{\Psi}$.

If $\epsilon \neq 0$, by construction $\hat{\Theta}^*$ is a model for the time series in $\hat{\Psi}$. To show that it is a *generating* system, since

$$det\, \hat{\Theta}^* = det\, \Gamma \cdot det\, \Theta^* = s \cdot det\, \Theta^*,$$

by theorem 6.1, $\hat{\Theta}^*$ has minimal degree, and by theorem 5.1, it is a generating system. □

7.2. Recursive update of a minimal-complexity controllable model. As mentioned earlier, all minimal complexity controllable models can be obtained from any generating system and one such controllable model. Thus we need to discuss the update of only a single minimal complexity controllable model. The idea is to arrange Θ^* in such a form that at each step of the update, a subset of its rows consitutes a minimal complexity controllable model. The procedure described in the previous sub-section will provide then a *simultaneous* update of *both* Θ^* and of a minimal complexity controllable model Θ^*_{contr}.

Because of definition (4.6), in order to keep track of minimality, it is necessary during the update to preserve the *row reducedness* of the generating system. Let the rows of Θ^* be denoted by θ_i:

$$\Theta^* =: \begin{pmatrix} \theta_1 \\ \vdots \\ \theta_q \end{pmatrix} \in \mathbf{R}^{q\times q}[s], \quad \theta_i \in \mathbf{R}^{1\times q}[s].$$

It will be assumed that Θ^* is row reduced with ordered row indices

$$\nu_i := deg\ \theta_i \leq deg\ \theta_{i+1} =: \nu_{i+1}, \quad i \in \underline{q-1}.$$

A consequence of this ordering of the row indices is that with Γ defined by (7.5), $\hat{\Theta}^* := \Gamma\Theta^*$, is also row reduced. It is necessary however to ensure the ordering of the indices during the update, as well. This depends on the error ϵ defined by (7.4). More precisely it holds if the index r is such that

$$\nu_r < \nu_{r+1}. \tag{7.6}$$

If not satisfied, ordering of the indices of the updated generating system can be achieved by premultiplying the rows of Θ^* having degree equal to ν_r, by an appropriate constant non-singular matrix Λ, so that only one among the corresponding ϵ_i's is non-zero. For details we refer to example **B**, at the end of the paper. These considerations are summarized in

PROPOSITION 7.2. *Let Θ^* be row reduced with ordered row indices as above, and let Γ be defined by (7.5). There exists $\Lambda \in \mathbf{R}^{q\times q}$, $det\,\Lambda \neq 0$, such that $\hat{\Theta}^* := \Gamma\Lambda\Theta^*$, is row reduced with ordered row indices.*

To describe the algorithm the index set $\mathbf{I}$ and the submatrix $\Theta_\mathbf{I}$ of Θ are introduced:

$$\mathbf{I} := \{i_1, \cdots, i_p\} \subset \underline{q}, \quad \Theta_\mathbf{I} := \begin{pmatrix} \theta_{i_1} \\ \vdots \\ \theta_{i_p} \end{pmatrix} \in \mathbf{R}^{p\times q}[s].$$

Algorithm: *Initialization step.*

$$\Theta^* := \Theta^*_{contr} := I_q, \quad \epsilon := \mathbf{w}_1^{\kappa_1}(0) = \mathbf{w}_1(-\kappa_1).$$

With Γ defined by (7.5), and through a row permutation we obtain

$$\hat{\Theta}^*(s) := \begin{pmatrix} I_{r-1} & & \\ & -\hat{\epsilon} & I_{q-r} \\ & s & \end{pmatrix} \in \mathbf{R}^{q\times q}[s], \quad \hat{\mathbf{I}} := \underline{q-1}, \quad \hat{\Theta}^*_{contr} = \hat{\Theta}^*_{\hat{\mathbf{I}}}. \tag{7.7}$$

Algorithm: *Step n.* Given Θ^*, let the error ϵ be defined by (7.4). First compute a pre-processing matrix Λ so that (7.6) is satisfied; define the new Θ^* and ϵ:

$$\Theta^* \leftarrow \Lambda\Theta^* \quad \text{and} \quad \epsilon \leftarrow \Lambda\epsilon.$$

Define Γ by (7.5), the index

$$k := min\ \{\ell \notin \mathbf{I} : \ \epsilon_\ell \neq 0\}, \tag{7.8}$$

and the updated index set $\hat{\mathbf{I}}$ as follows

$$\text{if} \quad r \notin \mathbf{I} \quad \Rightarrow \hat{\mathbf{I}} := \mathbf{I} \tag{7.9}$$

$$\text{if} \quad r \in \mathbf{I} \quad \Rightarrow \hat{\mathbf{I}} := \mathbf{I} \setminus \{r\} \cup \{k\}, \tag{7.10}$$

where if k does not exist, $\{k\}$ is the empty set. It follows that

$$\hat{\Theta}^* = \Gamma\Theta^* \quad \text{and} \quad \hat{\Theta}^*_{contr} = \hat{\Theta}^*_{\hat{\mathbf{I}}}. \tag{7.11}$$

Proof of step n. If (7.10) holds and the index k defined by (7.8) does not exist, $rank\,\Theta^*(0) = rank\,\hat{\Theta}^*(0) + 1$. In this case a controllable model of minimal complexity is obtained by eliminating the rth row. In all other cases $rank\,\Theta^*(0) = rank\,\hat{\Theta}^*(0)$, and if $r \in \mathbf{I}$, the rth row has to be replaced by the kth row. This proves the controllability of the updated model $\hat{\Theta}^*_{contr}$. Minimality follows by assumption of the recursive step, and definition (7.8). □

REMARK 7.1. From the above results we conclude that after one step, the complexity of the generating system increases and becomes

$$(0,\ n) \ \rightarrow \ (0,\ n+1);$$

that of the minimal complexity controllable systems increases as follows:

$$\text{if} \quad r \in \mathbf{I} \quad \text{and} \quad \nexists k : \quad (m,\ \sum_{i\in\mathbf{I}} \nu_i) \ \rightarrow \ (m+1,\ \sum_{i\in\mathbf{I}} \nu_i - \nu_r)$$

$$\text{otherwise}: \quad (m,\ \sum_{i\in\mathbf{I}} \nu_i) \ \rightarrow \ (m,\ \sum_{i\in\mathbf{I}} \nu_i - \nu_r + \nu_k)$$

□

8. Examples. A. In the scalar realization problem, a finite sequence of *Markov* parameters a_i, $0 \leq i \leq n$, is given. We seek to find all linear, time invariant models thereof. The data set $\mathbf{D}$ consists of the impulse response $\mathbf{w} : \mathbf{Z}_- \to \mathbf{R}^2$, defined as follows:

$$\mathbf{w}(-t) = 0,\ t > n;\ \ \mathbf{w}(-n) = \begin{pmatrix} 1 \\ a_0 \end{pmatrix};\ \ \mathbf{w}(-t) = \begin{pmatrix} 0 \\ a_{n-t} \end{pmatrix},\ \ n < t \leq 0.$$

Recursive modeling of $\mathbf{w}$ amounts to the succesive modeling of the shifted time series:

$$\sigma^{-n}\mathbf{w},\ \sigma^{-n+1}\mathbf{w},\ \sigma^{-n+2}\mathbf{w},\ \cdots,\ \sigma^{-1}\mathbf{w},\ \mathbf{w}.$$

Define the following sequence of polynomial matrices

$$\Gamma_i := \begin{pmatrix} s & 0 \\ -a_i & 1 \end{pmatrix},\ \ 0 \leq i \leq n.$$

It can be readily checked that the polynomial matrix Θ_i^* defined by the product

$$\Theta_i^* := \Gamma_i\Gamma_{i-1} \cdots \Gamma_1\Gamma_0 = \begin{pmatrix} s^{i+1} & 0 \\ -\pi & 1 \end{pmatrix},\ \ 0 \leq i \leq n,$$

where $\pi := a_i s^i + \cdots + a_1 s + a_0$, belongs to the modeling set $\mathcal{M}(\sigma^{-n+i}\mathbf{w})$, and has the invariant factors $(1,\ s^{i+1})$. Hence by theorem 6.1, Θ_i^* is a *generating system* for this model set. It follows from main theorem 5.1, that all other models are obtained as linear combinations of the two rows of Θ_i^*. For example, the input models are all multiples of s^{i+1}, the output models are given by all multiples of s^{j+1}, where j is the smallest subscript for which $a_j \neq 0$, and all input-output models are given by

$$q(\sigma^{-1})\mathbf{y}(t) = p(\sigma^{-1})\mathbf{u}(t) \quad \text{where} \quad [p \quad -q] = [a \quad b]\Theta_i^*,\ b \neq 0.$$

Moreover, the subset of controllable models is obtained by requiring that $b(0) \neq 0$. If, in addition, the complexity of the models is needed, then Θ_i^* needs to be in row reduced form. In this case, to proceed further, the numerical values of the Markov parameters are needed. Let for instance, $n = 4$ and

$$a_0 = 0,\ a_1 = 1,\ a_2 = 2,\ a_3 = 1,\ a_4 = 2.$$

A row reduced Θ_4^* is

$$\Theta_4^* = \begin{pmatrix} 2s^2 + s & s^2 - 1 \\ -3s^3 + s & 2s - 1 \end{pmatrix}.$$

Consequently, $(1,2)$ is the minimal complexity of the controllable models which are also input-output models; actually, in this case there is a unique such model, namely:

$$\mathbf{y}(t) \;=\; \mathbf{y}(t-2) \;+\; \mathbf{u}(t-1) \;+\; 2\mathbf{u}(t-2).$$

All controllable models of complexity $(1,3)$ are parametrized in terms of the real parameters α and β, where $\alpha+1 \neq 0$:

$$\begin{aligned}(\alpha+1)\mathbf{y}(t)+(\beta-2)\mathbf{y}(t-1)-\alpha\mathbf{y}(t-2)-\beta\mathbf{y}(t-3) \quad &= \\ (\alpha+1)\mathbf{u}(t-1) \;+\; (2\alpha+\beta)\mathbf{u}(t-2) \;+\; (2\beta-3)\mathbf{u}(t-3), \qquad & t\in\mathbf{Z}_-.\end{aligned}$$

□

B. This is a three-input, two-output realization example taken from Antoulas [1]. Let the following Markov parameters be provided:

$$A_0 \;=\; 0,\;\; A_1 \;=\; \begin{pmatrix} 1 & 1 & 1 \\ 2 & 2 & 0 \end{pmatrix},\;\; A_2 \;=\; \begin{pmatrix} 2 & 2 & 1 \\ 4 & 4 & 0 \end{pmatrix},$$

$$A_3 \;=\; \begin{pmatrix} 4 & 6 & 3 \\ 8 & 10 & 2 \end{pmatrix},\;\; A_4 \;=\; \begin{pmatrix} 8 & 11 & 6 \\ 16 & 21 & 7 \end{pmatrix}.$$

We are looking for all linear, time-invariant realizations of minimal complexity of these Markov parameters. We will solve this problem by applying the machinery developed above. The first step consists in transforming the Markov parameters into 3 time series (one for each input). This converts the problem to a tangential or directional realization problem. Let $(A_i)_j \in \mathbf{R}^2$, denote the jth column of the ith Markov parameter A_i, and $e_j \in \mathbf{R}^3$, the jth unit vector. The resulting data set is

$$\mathbf{D} \;:=\; \{\mathbf{w}_1,\; \mathbf{w}_2,\; \mathbf{w}_3\} \quad \text{where}$$

$$\mathbf{w}_j(-t)=0,\; t>4,\;\; \mathbf{w}_j(-4)=\begin{pmatrix} e_j \\ 0 \end{pmatrix},\;\; \mathbf{w}_j(-t)=\begin{pmatrix} 0 \\ (A_{4-t})_j \end{pmatrix},$$
$$t \;=\; 3,\; 2,\; 1,\; 0.$$

Recursive modeling of $\mathbf{D}$ amounts to succesive modeling of the shifted data set $\mathbf{D}_\sigma$, which is composed of the 15 time series

$$\sigma^{-4}\mathbf{w}_j,\;\; \sigma^{-3}\mathbf{w}_j,\;\; \sigma^{-2}\mathbf{w}_j,\;\; \sigma^{-1}\mathbf{w}_j,\;\; \mathbf{w}_j,\;\; j=1,\;2,\;3;$$

the sequence in which this modeling is carried out is irrelevant, as long as the shift invariance of the sets in (7.1) is satisfied.

Let Θ^* be a generating system for the set

$$\Psi \;=\; \{\sigma^{-t}\mathbf{w}_j,\; t=4,3,2,1,\; j=1,2,3\},$$

which corresponds to the first three Markov parameters A_1, A_2, A_3:

$$\Theta^* = \begin{bmatrix} -s & -s & s & s-1 & -1.5s+1 \\ 2s^2 & 2s^2 & 0 & 0 & s(2s-1) \\ 2s(s^2-1) & -2s & 0 & -2s^2 & -2s+1 \\ 0 & 0 & 0 & 0 & s^3 \\ 0 & 0 & 0 & s^3 & 0 \end{bmatrix}$$

The above generating system is in row reduced form; the first and third rows consitute a minimal complexity controllable model:

$$\mathbf{I} = \{1,3\},\ \Theta^*_{contr} = \Theta^*_{\mathbf{I}} = \begin{bmatrix} -s & -s & s & s-1 & -1.5s+1 \\ 2s(s^2-1) & -2s & 0 & -2s^2 & -2s+1 \end{bmatrix}.$$

We will now recursively model the remaining 3 time series as follows: first $\mathbf{w}_1$, then $\mathbf{w}_2$, and finally $\mathbf{w}_3$. Following (7.3), we have

$$\epsilon := \Theta^*(\sigma^{-1})\mathbf{w}_1(0) = [0 \quad 0 \quad -4 \quad 2 \quad 1]^t.$$

Since the three non-zero entries of ϵ correspond to rows of Θ^* having the same degree equal to 3, pre-processing is required in order to have *only one* of these components different from 0. The pre-processing matrix Λ and the corresponding Γ defined by (7.5), and the resulting updated generating system, are:

$$\Lambda = diag \left[I_2, \begin{pmatrix} 1 & 0 & 4 \\ 0 & 1 & -2 \\ 0 & 0 & 1 \end{pmatrix} \right], \quad \Gamma = diag\,[1,\ 1,\ 1,\ 1,\ s]$$

$$\Theta^* = \begin{bmatrix} -s & -s & s & s-1 & -1.5s+1 \\ 2s^2 & 2s^2 & 0 & 0 & s(2s-1) \\ 2s(s^2-1) & -2s & 0 & 4s^3-2s^2 & -2s+1 \\ 0 & 0 & 0 & -2s^3 & s^3 \\ 0 & 0 & 0 & s^4 & 0 \end{bmatrix}$$

Since $r = 5 \notin \mathbf{I}$, it follows from (7.9) that the set $\mathbf{I}$ remains the same after the update. Hence a controllable model of minimal complexity is given by the first and the third rows:

$$\Theta^*_{contr} = \Theta^*_{\mathbf{I}} = \begin{bmatrix} -s & -s & s & s-1 & -1.5s+1 \\ 2s(s^2-1) & -2s & 0 & 4s^3-2s^2 & -2s+1 \end{bmatrix}$$

The above procedure is repeated for the remaining two time series:

$$\epsilon := \Theta^*(\sigma^{-1})\mathbf{w}_2(0) = [1 \quad -2 \quad 1 \quad 0 \quad 0]^t.$$

Since the three non-zero entries of ϵ correspond to rows of the generating system having different degrees, no pre-processing is required. Therefore by (7.5):

$$\Gamma = diag\left[\begin{pmatrix} s & & \\ 2 & 1 & \\ -1 & & 1 \end{pmatrix}, \quad I_2\right]$$

$$\Theta^* = \begin{bmatrix} -s^2 & -s^2 & s^2 & s(s-1) & -1.5s^2+s \\ s(s-1) & s(s-1) & s & s-1 & (s-1)^2 \\ s(2s^2-1) & -s & -s & 4s^3-2s^2-s+1 & -.5s \\ 0 & 0 & 0 & -2s^3 & s^3 \\ 0 & 0 & 0 & s^4 & 0 \end{bmatrix}$$

In this case $r = 1 \in \mathbf{I}$ and hence k defined by (7.8) is needed. It turns out that $k = 2$ and consequently $\mathbf{I} = \{2, 3\}$:

$$\Theta^*_{contr} = \Theta^*_{\mathbf{I}} = \begin{bmatrix} s(s-1) & s(s-1) & s & s-1 & (s-1)^2 \\ s(2s^2-1) & -s & -s & 4s^3-2s^2-s+1 & -.5s \end{bmatrix}$$

At the final step we have

$$\epsilon = \Theta^*(\sigma^{-1})\mathbf{w}_3(0) = [0 \quad 0 \quad 4 \quad -2 \quad 0]^t.$$

Since the third and fourth rows have the same degree, pre-processing is required. The resulting quantities are

$$\Lambda = diag\left[I_2, \quad \begin{pmatrix} 1 & 2 \\ 0 & 1 \end{pmatrix}, \quad 1\right], \qquad \Gamma = diag\,[1, \ 1, \ 1, \ s, \ 1]$$

$$\Theta^* = \begin{bmatrix} -s^2 & -s^2 & s^2 & s(s-1) & -1.5s^2+s \\ s(s-1) & s(s-1) & s & s-1 & (s-1)^2 \\ s(2s^2-1) & -s & -s & -2s^2-s+1 & 2s^3-.5s \\ 0 & 0 & 0 & -2s^4 & s^4 \\ 0 & 0 & 0 & s^4 & 0 \end{bmatrix}$$

Since $r = 4$, $\mathbf{I} = \{2, 3\}$. Hence one controllable model of minimal complexity is given by the second and the third rows. A parametrization of all such controllable models is given by adding an arbitrary constant multiple of the first row to the second, and an arbitrary first-order polynomial multiple of the first row to the third:

$$\Theta^*_{contr}(s; \alpha, \beta, \gamma) = \begin{bmatrix} \alpha & 1 & 0 & 0 & 0 \\ \beta s + \gamma & 0 & 1 & 0 & 0 \end{bmatrix} \Theta^*,$$

where all values of the parameters α, β, γ are admissible. The resulting minimal complexity controllable models are input-output models as well. Hence, after partitioning Θ^*_{contr} as in (4.1) with $g = p = 2$, $q = 5$, the transfer function of these models is given by $Z = T^{-1}Q$.

We conclude this example by mentioning that the input models are multiples of $s^5 I_3$, and the output models are multiples of $s^4 I_2$. □

REFERENCES

[1] A.C. Antoulas, On recursiveness and related topics in linear systems, *IEEE Transactions on Automatic Control*, **AC-31**: 1121-1135 (1986).

[2] A.C. Antoulas, J.A. Ball, J. Kang, and J.C. Willems, On the the solution of the minimal rational interpolation problem, *Linear Algebra and its Applications*, **137/138**: 511-573 (1990).

[3] A.C. Antoulas and J.C. Willems, A behavioral approach to linear exact modeling, *IEEE Transactions on Automatic Control*, **AC-38** (1993).

[4] P.A. Fuhrmann, Algebraic methods in system theory, in **Mathematical System Theory** (A.C. Antoulas Editor), pp. 233-265, Springer Verlag (1991).

[5] I. Gohberg, M.A. Kaashoek, and L. Lerer, A directional partial realization problem, **Systems and Control Letters, 17**: 305-314 (1991).

[6] T. Kailath, **Linear Systems**, Prentice-Hall, (1980).

[7] J.C. Willems, Paradigms and puzzles in the theory of dynamical systems, *IEEE Transactions on Automatic Control*, **AC-36**: 259-294 (1991).

POLE PLACEMENT, INTERNAL STABILIZATION AND INTERPOLATION CONDITIONS FOR RATIONAL MATRIX FUNCTIONS: A GRASSMANNIAN FORMULATION

JOSEPH A. BALL * AND JOACHIM ROSENTHAL†

Abstract. The problem of pole placement via dynamic feedback and the bitangential interpolation problem are shown to be both particular instances of a general subspace interpolation problem formulated for rational curves from projective space into a Grassmannian manifold. The problem of determining the minimal degree d for an interpolant in terms of the problem data is shown to be computable via intersection theory in projective space. Using the projective dimension theorem bounds for the minimal degree interpolant curve are given.

Key words. Multivariable Systems, Interpolation Problems, Dynamic Feedback Compensation, Autoregressive Systems.

AMS(MOS) subject classifications. 32E30, 65D05, 93B55, 93C35, 14M15.

1. Introduction. The problem of pole placement via dynamic output feedback is as follows. One is given a plant P described in the frequency domain through the input-output relation:

$$\hat{y}(s) = G(s)\,\hat{u}(s). \tag{1.1}$$

One seeks to build a compensator C characterized through the relation:

$$\hat{u}(s) = F(s)\,\hat{y}(s) + \hat{v}(s) \tag{1.2}$$

such that the resulting closed loop transfer function

$$G_F(s) := (I - G(s)F(s))^{-1}G(s) \tag{1.3}$$

is well defined and has poles in certain prespecified locations $\{s_1, s_2, \ldots, s_\ell\}$.

In order to describe the poles of the closed loop transfer function we introduce a left coprime factorization of $G(s)$ and a right coprime factorization of $F(s)$:

$$G(s) = D_{LG}^{-1}(s)N_{LG}(s) \qquad F(s) = N_{RF}(s)D_{RF}^{-1}(s). \tag{1.4}$$

Using those coprime factorizations it is straight forward to show (compare with [20]) that the poles of $G_F(s)$ are zeroes of the polynomial

$$\det(D_{LG}(s)D_{RF}(s) - N_{LG}(s)N_{RF}(s)). \tag{1.5}$$

---

* Department of Mathematics, Virginia Tech, Blacksburg, VA 24061. The work of the first author was supported in part by NSF grant DMS-9101400.

† Department of Mathematics, University of Notre Dame, Notre Dame, IN 46556. The work of the second author was supported in part by NSF grant DMS-9201263.

Using this characterization one has therefore the following interpolation condition:

$$\text{(1.6)}\ \det(D_{LG}(s_k)D_{RF}(s_k) - N_{LG}(s_k)N_{RF}(s_k)) = 0 \ \text{ for } \ k = 1, \ldots, \ell.$$

Since the early work of Brash and Pearson [10] there has been done much work on determining when one can achieve the placement of the poles with a static compensator, or, when this is not possible, what is the minimal possible McMillan degree d (at least generically) of a compensator for which the pole placement is achieved. (Compare with [20,23], where also more references can be found.)

Another constraint often sought in feedback design is that there be no pole-zero cancellation between plant and compensator in some region of the complex plane C_g. If the plant is stable and we use the closed loop transfer function $S(s) = (I - G(s)F(s))^{-1}$ as the design parameter rather than $F(s)$ then one requires that $S(s)$ be analytic on C_g and, in addition, at points of C_g where $G(s)$ is analytic and $G(s)$ has simple zeros, $S(s)$ in addition must satisfy left tangential interpolation conditions of the sort

$$x_k S(s_k) = y_k, \quad \text{for } \ k = 1, \ldots, \ell, \tag{1.7}$$

say, where x_k and y_k are row vectors of appropriate size. On the other hand, at points in C_g where the plant $P(s)$ has poles but not zeros and the poles of $G(s)$ are simple, one gets right interpolation conditions on $S(s)$ of the form

$$S(s_i)u_i = v_i \ \text{ for } \ i = 1, \ldots, \ell, \tag{1.8}$$

where u_i and v_i are column vectors of appropriate size. Here the data $\{x_i, y_i, u_i, v_i, s_i\}$ are all computable from a minimal realization of the original plant P; see Chapters 23 and 25 of [4] and [5] for elementary discussions and [7,8] for a very general formulation. In the context of H^∞-control one requires in addition that a norm constraint

$$||S(s)|| \leq \gamma \tag{1.9}$$

on S hold for all $s \in C_g$. In the context of modeling time series with minimal degree of complexity (see e.g. [2]), the problem of finding the minimal McMillan degree of an interpolant comes up also for this type of interpolation problem. The minimal degree issue for this problem also has a long history; recent treatments of the general matrix case are [1] and [6].

2. A geometric reformulation. It is the purpose of this article to formulate a more general interpolation problem which contains the problems (1.6), (1.7) and (1.8) formulated in the introduction as particular instances. In order to achieve this goal we reformulate the above problems in the sequel in a geometric form.

2.1. Pole placement. We assume that the behavior of the plant P is presented as the set of input-output pairs which satisfy a linear system of homogeneous differential equations as proposed in [24]. For this consider a $m \times (m+p)$ matrix $P(s) = (f_{ij}(s))$ whose entries are elements of the ring $\mathbb{C}[s]$. $P(s)$ induces a system of *autoregressive equations* given by:

$$P(\frac{d}{dt})\begin{pmatrix} x(t) \\ y(t) \end{pmatrix} = 0. \tag{2.1}$$

A feedback compensator (compare with [20,21,25]) in this framework is a $p\times(m+p)$ polynomial matrix $C(s)$. The dynamic of the closed loop system is then characterized through the autonomous system

$$\begin{pmatrix} P(\frac{d}{dt}) \\ C(\frac{d}{dt}) \end{pmatrix} \cdot \begin{pmatrix} x(t) \\ y(t) \end{pmatrix} = 0. \tag{2.2}$$

We call a compensator $C(s)$ *admissible* (compare with [21,25]) if the characteristic polynomial

$$\varphi(s) := \det \begin{pmatrix} P(s) \\ C(s) \end{pmatrix} \not\equiv 0. \tag{2.3}$$

The degree of $C(s)$ is defined as the highest degree of all full size minors of $C(s)$. (Compare with [12,18,19,21,24,25].) The pole placement problem can now be formulated in the following way: Given a $m\times(m+p)$ polynomial matrix $P(s)$ of degree n and a set of numbers (poles) $s_1, \ldots, s_\ell$. Construct an admissible $p \times (m+p)$ polynomial matrix $C(s)$ such that

$$\text{row space } (P(s_k)) \cap \text{row space } (C(s_k))) \neq \{0\} \tag{2.4}$$

for $k = s_1, \ldots, s_\ell$. If possible find the minimal number $d = d(m,p,n)$ which guarantees the existence of a degree d interpolant $C(s)$ for any given set of data. For the connection between the formulation (1.6) and (2.4) we refer to [20,23].

2.2. Left tangential interpolation conditions. We now consider the problem involving left tangential interpolation conditions of the form (1.7). We suppose that $S(s)$ has a right coprime factorization over $\{s_1, \ldots, s_\ell\}$ of the form $S(s) = N_R(s)D_R(s)^{-1}$. Then the left tangential interpolation conditions can be expressed as

$$\begin{pmatrix} x_k & -y_k \end{pmatrix} \begin{pmatrix} N_R(s_k) \\ D_R(s_k) \end{pmatrix} = 0. \tag{2.5}$$

for $k = 1, \ldots, \ell$. In terms of a left coprime factorization $S(s) = D_L(s)^{-1}$ $N_L(s)$ these conditions are equivalent to: Construct a $p\times(m+p)$ polynomial matrix $(D_L(s)\, N_L(s))$ such that

$$\text{span} \begin{pmatrix} x_k & y_k \end{pmatrix} \subseteq \text{row space} \begin{pmatrix} D_L(s_k) & N_L(s_k) \end{pmatrix} \tag{2.6}$$

for $k = 1, \ldots, \ell$. Again we are interested in the minimal degree d interpolant $(D_L(s)\, N_L(s))$. Note that not all solutions $(D_L(s)\, N_L(s))$ of (2.6) have the property that $D_L(s)$ is invertible at infinity. Still every solution represents an autoregressive system. (Compare with [12,16,19,24].)

2.3. Right tangential interpolation conditions. Next we consider right tangential interpolation conditions of the form (1.8). Again we assume that $S(s)$ has a left coprime factorization $S(s) = D_L(s)^{-1} N_L(s)$ over $\{s_1, \ldots, s_\ell\}$. Then the interpolation conditions (1.8) can be equivalently expressed in terms of the left coprime factors D_L and N_L as

$$(N_L(s_i)\; D_L(s_i)) \begin{pmatrix} u_i \\ -v_i \end{pmatrix} = 0. \tag{2.7}$$

For $i = 1, \ldots, \ell$, let V_i be the n-dimensional subspace of $\mathbb{C}^n$ given by

$$V_i = \{x \in \mathbb{C}^{1\times n} : x \begin{pmatrix} u_i \\ -v_i \end{pmatrix} = 0\}. \tag{2.8}$$

Then the conditions (1.8) take the form: Construct a $p \times (m+p)$ polynomial matrix $(N_L(s)\, D_L(s))$ such that

$$\text{row space } \begin{pmatrix} N_L(s_i) & D_L(s_i) \end{pmatrix} \subseteq V_i \tag{2.9}$$

for $i = 1, \ldots, s_\ell$.

At this point we want to mention that problems (2.4), (2.6) and (2.9) all require the construction of a holomorphic map (Hermann Martin curve) having domain the Riemann sphere and range a fixed Grassmann manifold. In addition, for many applications we seek a polynomial matrix $(N_L(s)\, D_L(s))$ which has in addition the property that $D_L(s)$ is invertible. In this paper we will not consider these genericity questions. For techniques which resolve these genericity questions in the case of the pole placement problem we refer to [20,23].

In the next section we formulate a general interpolation problem which includes the above problems as special cases.

3. A general interpolation problem of Schubert type. Before stating the general interpolation problem which unifies the three types of interpolation conditions (2.4), (2.6) and (2.9) given above, we need some preliminaries.

Denote with $\mathbb{P}^1 = \mathbb{P}(\mathbb{C})$ the complex projective line, i.e. the Riemann sphere and let $G_p(\mathbb{C}^n)$ be the Grassmann variety consisting of all p-dimensional subspaces of n-dimensional complex Euclidean space $\mathbb{C}^n$, where we assume that $n \geq p$ and the elements of $\mathbb{C}^n$ are considered as row vectors. The Plücker embedding $\mathcal{P}$ of $G_p(\mathbb{C}^n)$ into $\mathbb{P}(\wedge^p \mathbb{C}^n) = \mathbb{P}^{\binom{n}{p}-1}$ is defined as follows. If $W \in G_p(\mathbb{C}^n)$, choose a basis $\{w_1, \ldots, w_p\}$ for W and define

$$\begin{aligned} \mathcal{P}: \quad G_p(\mathbb{C}^n) &\longrightarrow \mathbb{P}(\wedge^p \mathbb{C}^n) \\ \operatorname{span}(w_1, \ldots, w_p) &\longmapsto w_1 \wedge \ldots \wedge w_p. \end{aligned} \tag{3.1}$$

In terms of coordinates let $e_1, \ldots, e_n$ be a basis for $\mathbb{C}^n$ and suppose the basis vectors $w_1, \ldots, w_p$ for W are expressed as the row vectors $w_1 = \begin{pmatrix} a_{11} & \ldots & a_{1n} \end{pmatrix}$, $w_2 = \begin{pmatrix} a_{21} & \ldots & a_{2n} \end{pmatrix}$, …, $w_p = (a_{p1}, \ldots, a_{pn})$ with respect to this basis for $\mathbb{C}^n$. Then a basis for $\wedge^p(\mathbb{C}^n)$ consists of the $\begin{pmatrix} n \\ p \end{pmatrix}$ elements $e_{\underline{i}} = e_{i_1} \wedge e_{i_2} \wedge \ldots \wedge e_{i_p}$ where $\underline{i} = \{i_1, \ldots, i_p\}$ sweeps all p-tuples of indices subject to $i_1 < i_2 < \cdots < i_p$. With respect to this basis for $\wedge^p \mathbb{C}^n$, the homogeneous coordinates of $\mathcal{P}(W)$ in $\mathbb{P}(\wedge^p \mathbb{C}^n)$ are given by

$$\mathcal{P}(W) = \sum_{\underline{i}} a_{\underline{i}} e_{\underline{i}} \tag{3.2}$$

where $a_{\underline{i}}$ is the p-minor of the matrix

$$A = \begin{pmatrix} a_{11} & a_{12} & \ldots & a_{1n} \\ a_{21} & a_{22} & \ldots & a_{2n} \\ \vdots & \vdots & & \vdots \\ a_{p1} & a_{p2} & \ldots & a_{pn} \end{pmatrix} \tag{3.3}$$

formed by the $i_1 th, i_2 th, \ldots, i_p th$ columns. The image of $G_p(\mathbb{C}^n)$ in $\mathbb{P}(\wedge^p \mathbb{C}^n)$ under the Plücker embedding is characterized through the so called Shuffle relations, a set of homogeneous quadratic equations:

$$\sum_{\lambda}^{n} (-1)^{\lambda} a_{i_1, \ldots, i_{p-1}, j_\lambda} \cdot a_{j_1, \ldots, \hat{j}_\lambda, \ldots, j_{p+1}} = 0 \tag{3.4}$$

where $i_1, \ldots, i_{p-1}$ and $j_1, \ldots, j_{p+1}$ are any sequences of integers with

$$1 \le i_\alpha, j_\beta \le n, \tag{3.5}$$

and the symbol $\wedge$ means that j_λ is removed and

$$a_{i_1, i_2, \ldots, i_p} = (\operatorname{sign}\sigma) a_{\sigma(i_1, i_2, \ldots, i_p)} \tag{3.6}$$

for any permutation σ. In particular we see that the image of $G_p(\mathbb{C}^n)$ in $\mathbb{P}(\wedge^p \mathbb{C}^n)$ is defined by multilinear nonlinear relations on the Plücker coordinates.

By a flag $\mathcal{F}$ we mean a collection $\{V_i : i = 1, \ldots, n\}$ of nested subspaces of $\mathbb{C}^n$ such that

1. dim $V_i = i$
2. $\{0\} \subset V_1 \subset \ldots \subset V_n = \mathbb{C}^n$. For $\underline{i} = \{i_1, \ldots, i_p\}$ a p-tuple of integers satisfying $1 \le i_1 < i_2 < \ldots < i_p \le n$

Associated to each flag $\mathcal{F}$ is a Schubert cell defined through

$$\mathcal{C}(\underline{i},\mathcal{F}) := \{W \in G_p(\mathbb{C}^n) : \dim(W \bigcap V_{i_j}) = j \text{ for } j = 1,\ldots,p\}. \tag{3.7}$$

The Zariski closure of $\mathcal{C}(\underline{i},\mathcal{F})$ in $G_p(\mathbb{C}^n)$ is given by the Schubert variety

$$\mathcal{S}(\underline{i},\mathcal{F}) := \{W \in G_p(\mathbb{C}^n) : \dim\, W \cap V_{i_j} \geq j \text{ for } j = 1,\ldots,p\}. \tag{3.8}$$

In order to characterize $\mathcal{C}(\underline{i},\mathcal{F})$ and $\mathcal{S}(\underline{i},\mathcal{F})$ in terms of Plücker coordinates choose a basis $e_1,\ldots,e_n$ of $\mathbb{C}^n$ such that $\text{span}\{e_1,\ldots,e_k\} = V_k$, $k = 1,\ldots,n$. Using this basis it is not difficult to show that all elements of $\mathcal{C}(\underline{i},\mathcal{F})$ have pivot indices $\underline{i} = (i_1,\ldots,i_p)$ in the row reduced echelon form. In particular $\mathcal{C}(\underline{i},\mathcal{F})$ is a cell, i.e. isomorphic to $\mathbb{C}^N$ where $N := \sum_{\nu=1}^{p} i_\nu - \nu$ is the dimension of the cell $\mathcal{C}(\underline{i},\mathcal{F})$ which is also the dimension of $\mathcal{S}(\underline{i},\mathcal{F})$.

In order to describe the defining equations of $\mathcal{S}(\underline{i},\mathcal{F})$ in $\mathbb{P}(\wedge^p\mathbb{C}^n)$ we introduce a partial order for the Plücker coordinates. Denote with $\underline{i} = (i_1,\ldots,i_p)$ and $\underline{j} = (j_1,\ldots,j_p)$ and define:

$$\underline{i} \leq \underline{j} \iff i_s \leq j_s \;\forall\, s \tag{3.9}$$

Using this partial order one has the classical result: (Compare with [13,17])

LEMMA 3.1. *$\mathcal{S}(\underline{i},\mathcal{F})$ is a projective variety of dimension $\sum_{\nu=1}^{p} i_\nu - \nu$ and defined through the relations:*

$$\mathcal{S}(\underline{i},\mathcal{F}) = \{x \in G_p(\mathbb{C}^n) \subseteq \mathbb{P}(\wedge^p\mathbb{C}^n) : x_{\underline{j}} = 0 \;\; for \;\; \underline{j} \not\leq \underline{i}\} \tag{3.10}$$

We are now ready to formulate the interpolation problem. We suppose that we are given ℓ points $s_1,\ldots,s_\ell$ in $\mathbb{P}^1$ together with ℓ Schubert varieties $\mathcal{S}_k = \mathcal{S}(\underline{i}_k,\mathcal{F}_k)$ $(k = 1,\ldots,\ell)$. The interpolation problem then is: *Find all $p\times n$ full rank polynomial matrices $F(s)$ such that there are points $Q_k \in \mathcal{S}_k$ with*

$$\text{row space } F(s_k) \subset Q_k \in \mathcal{S}_k \tag{3.11}$$

for $k = 1,\ldots,\ell$. Moreover find a polynomial matrix of minimal degree d which satisfies the interpolation condition (3.11). Using a more algebraic geometric language we seek a rational curve (Hermann-Martin curve) $F : \mathbb{P}^1 \to G_p(\mathbb{C}^n)$ of minimal degree which satisfies condition (3.11).

To recover the classical interpolation conditions, we should also insist that the value $F(s_k)$ of F at s_k is full rank for each interpolation node s_k, i.e. row space $F(s_k) \in \mathcal{S}_k$ or the Herman-Martin curve $F : \mathbb{P}^1 \to G_p(\mathbb{C}^n)$ is regular at s_k. However in the discussion to follow in the next section we will not impose this requirement. In practice this also is a genericity requirement which can be added later.

We next show how interpolation conditions of the form (3.11) encompass each of the sets of interpolation conditions (2.4), (2.6) and (2.9) as special cases.

To see how (2.4) is a special case of (3.11), take the p-tuples $\underline{i}_k = \{i_1^{(k)}, \ldots, i_p^{(k)}\}$ of indices all to be of the form $i_1^{(k)} = m, i_2^{(k)} = m+2, i_3^{(k)} = m+3, \ldots, i_p^{(k)} = m+p = n$ while $\mathcal{F}_k = \{V_1^{(k)}, \ldots, V_n^{(k)}\}$ is taken to be any flag such that $V_m^{(k)} =$ row space $P(s_k)$ for $k = 1, \ldots, \ell$.

To see how the interpolation conditions (2.6) assume the form (3.11) consider the p-tuples $\underline{i}_k = \{i_1^{(k)}, \ldots, i_p^{(k)}\}$ all such that $i_1^{(k)} = 1$, $i_2^{(k)} = m+2, i_3^{(k)} = m+3, \ldots, i_p^{(k)} = m+p = n$ and $\mathcal{F}_k = \{V_1^{(k)}, \ldots, V_n^{(k)}\}$ is any flag such that $V_1^{(k)} = \text{span}\begin{pmatrix} x_k & y_k \end{pmatrix}$. It is then immediate that condition (2.4) is equivalent to the condition that row space$(D_L(s_k)N_L(s_k)) \in \mathcal{S}_k(\underline{i}, \mathcal{F})$.

We remark that a similar Grassmannian formulation of tangential interpolation conditions appears in [3].

Finally note that conditions (2.9) just say that the Hermann Martin curve described by row space$(N_L(s)D_L(s))$ is contained at s_i in the sub-Grassmann manifold $\mathcal{S}_i := \{W \in G_p(\mathbb{C}^n) : W \subset V_i\}$ of $G_p(\mathbb{C}^n)$ which is clearly a Schubert variety as well. In other words condition (2.9) is also of the form (3.11).

4. Interpolants of minimal degree. In this section we consider the problem of determining the minimal degree d for a solution F of the interpolation problem (3.11) described in the previous section.

We first recall some definitions. Consider the set of all $p \times (m+p)$ polynomial matrices of full row rank. An equivalence class of row equivalent matrices is called an autoregressive system. (Compare with [18,19].) As done in [12,19] we define the degree of an autoregressive system as the highest degree of the set of full size minors. Using notation from [19] we denote with $A_{p,m}^{\prec d}$ the set of all autoregressive systems of degree at most d. For every autoregressive system $F(s) \in A_{p,m}^{\prec d}$ choose a basis $\{w_1(s), \ldots, w_p(s)\}$ of the row space of $F(s)$. One has a well defined map

$$\begin{array}{cccc} (4.1) & \mathcal{P}_d : & A_{p,m}^{\prec d} & \longrightarrow & \mathbb{P}(\mathbb{C}^{d+1} \otimes \wedge^p \mathbb{C}^{m+p}) \\ & & \text{span}(w_1(s), \ldots, w_p(s)) & \longmapsto & w_1(s) \wedge \ldots \wedge w_p(s). \end{array}$$

Note that for $d = 0$, $\mathcal{P}_d$ is exactly the Plücker embedding (3.1). As shown in [20] the image of $A_{p,m}^{\prec d}$ in $\mathbb{P}(\mathbb{C}^{d+1} \otimes \wedge^p \mathbb{C}^{m+p})$ is a projective variety of dimension $d(m+p) + mp$, which we like to denote with $K_{p,m}^d$. In order to describe each interpolation condition in (3.11) we introduce the following subset of $A_{p,m}^{\prec d}$:

$$(4.2) \quad \sigma(\mathcal{S}_k, s_k, d) := \{F(s) \in A_{p,m}^{\prec d} : \text{ row space } F(s_k) \in \mathcal{S}_k = \mathcal{S}(\underline{i}_k, \mathcal{F}_k)\}.$$

From this definition it is obvious that the interpolation problem (3.11) has a solution in the class of interpolants of degree d if and only if

$$(4.3) \qquad \bigcap_{k=1}^{\ell} \sigma(\mathcal{S}_k, s_k, d) \neq \emptyset.$$

The following Lemma shows that the intersection can be done in the variety $K^d_{p,m}$.

LEMMA 4.1. *If* $P(s) \in \sigma(\mathcal{S}_k, s_k, d)$ *then the whole fibre* $\mathcal{P}_d^{-1}(\mathcal{P}_d(P(s)))$ $\subseteq \sigma(\mathcal{S}_k, s_k, d)$.

Proof. If $\tilde{P}(s) \in \mathcal{P}_d^{-1}(\mathcal{P}_d(P(s)))$ then $P(s)$ and $\tilde{P}(s)$ are H-equivalent (see [20] for a definition). In particular at the point s_k one has $\operatorname{rowsp} P(s_k) = \operatorname{rowsp} \tilde{P}(s_k)$. □

The following theorem describes the image of the set $\sigma(\mathcal{S}_k, s_k, d)$ under the map $\mathcal{P}_d$.

THEOREM 4.2. $\mathcal{P}_d(\sigma(\mathcal{S}_k, s_k, d)) \subseteq \mathbb{P}(\mathbb{C}^{d+1} \otimes \wedge^p \mathbb{C}^{m+p})$ *is a projective variety of dimension* $d(m+p) + \sum_{\nu=1}^{p} i_\nu - \nu$. *The defining equations are given by a set of quadratic equations describing the set* $K^d_{p,m} \subseteq \mathbb{P}(\mathbb{C}^{d+1} \otimes \wedge^p \mathbb{C}^{m+p})$ *plus a set of linear equations coming from the requirement that* $f_{\underline{j}}(s_k) = 0$ *for all* $\underline{j} \not\leq \underline{i}$, *where* $f_{\underline{j}}(s_k)$ *is the* $\underline{j}$*-th Plücker coordinate of* $F(s_k)$.

Proof. Without loss of generality we can choose a basis $\{e_1, \ldots, e_{m+p}\}$ of $\mathbb{C}^{m+p}$ such that the flag $\mathcal{F}$ is given by

$$\{0\} \subset \operatorname{span}\{e_1\} \subset \ldots \subset \operatorname{span}\{e_1, \ldots, e_{m+p}\}. \tag{4.4}$$

From [20] it is known that $K^d_{p,m} \subseteq \mathbb{P}(\mathbb{C}^{d+1} \otimes \wedge^p \mathbb{C}^{m+p})$ is described by a set of quadratic equations. From the description of the Schubert variety $\mathcal{S}(\underline{i}, \mathcal{F})$ in Lemma (3.1) it is clear that the equations $f_{\underline{j}}(s_k) = 0$ for $\underline{j} \not\leq \underline{i}$ are necessary and sufficient. Moreover these equations are linear in the coordinates of $\mathbb{P}(\mathbb{C}^{d+1} \otimes \wedge^p \mathbb{C}^{m+p})$.

In order to prove the dimension formula we assume without loss of generality that $s_k = \infty$. Furthermore assume that $F(s) \in \mathcal{P}_d(\sigma(\mathcal{S}_k, s_k, d))$ is represented in canonical row proper form (see [9, page 333] for a definition). Counting all free parameters one sees that the "high coefficient matrix" of $F(s)$ has $\sum_{\nu=1}^{p} i_\nu - \nu$ free parameters. In addition one has $d(m+p)$ other free parameters giving the stated dimension formula. □

Combining Lemma (4.1) and Theorem (4.2) one has the following main result:

THEOREM 4.3. *The interpolation problem* (3.11) *has a solution in the class of degree d interpolants if and only if the algebraic set*

$$\bigcap_{k=1}^{\ell} \mathcal{P}_d(\sigma(\mathcal{S}_k, s_k, d)) \tag{4.5}$$

is nonempty.

We conclude this section with several examples:

4.1. Pole placement. In this case the Schubert varieties $\mathcal{S}_k = \mathcal{S}(\underline{i}_k, \mathcal{F}_k)$ in problem (3.11) are Schubert hypersurfaces, i.e. varieties of codimension 1 in $G_p(\mathbb{C}^{m+p})$. From Theorem (4.2) then follows that

$\mathcal{P}_d(\sigma(\mathcal{S}_k, s_k, d))$ is a codimension one variety of $K^d_{p,m}$ as well, described by one linear equation in $\mathbb{P}(\mathbb{C}^{d+1} \otimes \wedge^p \mathbb{C}^{m+p})$. Using this fact together with the projective dimension theorem the second author [20] concluded that (2.4) has always a solution if and only if $\dim K^d_{p,m} = d(m+p) + mp \geq \ell$. Moreover generically a solution is admissible and describes a proper transfer function. If $d = 0$ those results reduce to the well known pole placement results of Brockett and Byrnes [11].

4.2. Degree 0 interpolation. If one is interested in interpolation solutions of (3.11) which have degree $d = 0$ (the static case) one deals with the classical Schubert intersection problem. Indeed in this case $A^{\prec d}_{p,m} = A^0_{p,m} = G_p(\mathbb{C}^{m+p})$ and $\sigma(\mathcal{S}_k, s_k, d) = \mathcal{S}(\underline{i}_k, \mathcal{F}_k)$. The algebraic set described in (4.5) has then a solution as soon as the product of the corresponding Schubert cocycles in $H^*(G_p(\mathbb{C}^n), \mathcal{Z})$ is nonzero. (Compare with [13].)

4.3. The single output case. Assume problem (3.11) is given with $p = 1$. In this case the variety $K^d_{p,m} = K^d_{1,m}$ is equal to $\mathbb{P}(\mathbb{C}^{d+1} \otimes \mathbb{C}^{m+1}) = \mathbb{P}^{dm+d+m}$. Moreover the variety $\mathcal{P}_d(\sigma(\mathcal{S}_k, s_k, d))$ is a linear subspace of $\mathbb{P}^{dm+d+m}$. Therefore (4.5) is in this case a purely linear problem and a solution is always guaranteed (over any field) as soon as

$$dm + d + m \geq \sum_{k=1}^{\ell} (m + 1 - i_1^{(k)}). \tag{4.6}$$

In particular if one is only interested in right tangential interpolation conditions of type (2.9), (i.e. $i_1^{(k)} = m$ for $k = 1, \ldots, \ell$) one gets the sufficient condition

$$dm + d + m \geq \ell \tag{4.7}$$

which is also necessary for the generic set of interpolation data.

REFERENCES

[1] A.C.Antoulas, J.A.Ball J.Kang(Kim) and J.C.Willems, *On the solution of the minimal interpolation problem,* Linear Alg. and Appl., 137/138 (1990), 511–573.

[2] A.C.Antoulas and J.C.Willems, *A basis-free approach to linear exact modeling,* preprint.

[3] J.A.Ball and J.W.Helton, *Lie groups over the field of rational functions, signed spectral factorization, signed interpolation and amplifier design,* J. Operator Theory, 9 (1982), 1964.

[4] J.A.Ball, I.Gohberg and L.Rodman, *Interpolation of Rational Matrix Functions,* Birkhäuser-Verlag, Basel-Berlin-Boston, 1990.

[5] J.A.Ball, I.Gohberg and L.Rodman, "Sensitivity minimization and bitangential Nevanlinna-Pick interpolation in contour integral form," in *Signal Processing II: Control Theory and Applications* (ed. F.A. Grünbaum et al), IMA Vol. in Math. and its Appl. Vol. 23, Springer-Verlag, (1990), pp. 3–35.

FEEDBACK STABILIZIBILITY OVER COMMUTATIVE RINGS

J.W. BREWER*, L.C. KLINGLER*, AND WILAND SCHMALE†

Key words. linear systems, pole assignability, dynamic feedback

AMS(MOS) subject classifications. 93b55

In this paper, we survey a part of feedback stabilization for systems over commutative rings. Since many excellent sources exist which describe the motivation behind the study of systems over rings, we shall not touch on that here. The interested reader is referred to [1], [8], [9], [12], [13], and [14].

Let us make only one point. As is typical in such situations, adding some axiomatic generality serves to bring the essential ideas into focus. Thus, the treatment here for commutative rings helps to clarify the situation for fields.

By a *system* over a commutative ring R , we mean a pair (A, B) of matrices over R. The system is called n-dimensional if the matrix A is $n \times n$ and it will be our convention that the matrix B will be of size $n \times m$. With this convention we just have a sequence

$$R^m \xrightarrow{B} R^n \xrightarrow{A} R^n$$

The system is called *reachable* if and only if the R-module generated by the columns of the matrix

$$\begin{bmatrix} B & AB & \dots & A^{n-1}B \end{bmatrix}$$

equals R^n. In terms of homomorphisms, this means that

$$imB + imAB + \dots = R^n$$

where "imF" denotes the image of the homomorphism F.

We are interested initially in a very strong form of feedback stabilization for a ring R which is valid, for example, for fields. Now, it is an interesting exercise in linear algebra to prove the following:

PROPOSITION 1.1. *Suppose that (A, B) is an n-dimensional system over a commutative ring R. Suppose further that for each set $\{r_1, r_2, \dots r_n\}$ of elements of R, there exists a matrix K, depending on $r_1, r_2, \dots, r_n$, such*

* Department of Mathematics, Florida Atlantic University, Boca Raton, Florida 33431. The first named author was supported in part by the Institute for Mathematics and Its Applications with funds provided by the National Science Foundation and the second named author was partially supported by an NSF grant.

† Department of Mathematics, Carl v. Ossietzky Universitat Oldenburg, W-2900 Oldenburg, Germany.

that the characteristic polynomial of $A+BK = (X-r_1)(X-r_2)\dots(X-r_n)$. *Then the system is reachable.*

This fact was first noted by E.Sontag in [12]. In words, it can be phrased as follows: If the system (A, B) is *pole assignable*, then the system is reachable. Thus, the only kinds of systems whose poles can be placed arbitrarily are the reachable ones. This means that given a ring R, which we would like to verify as having the desired stabilization property, we have only to check that the reachable systems over it can have their poles assigned arbitrarily.

Before we can define our strong form of stabilizibility, we need an additional piece of motivation. Essentially, the following result says that for a single-input reachable system you have, via state feedback, complete control over its characteristic polynomial. We require some terminology and all of this can be found in [1].

Let (A, B) be a system over a ring R. If B consists of a single vector b, then the system is called *single-input* and it is customary to write (A, b) in place of (A, B). Note that if the system is n-dimensional, then (A, b) is reachable if and only if the determinant of the matrix

$$\begin{bmatrix} b & Ab & \dots & A^{n-1}b \end{bmatrix}$$

is a unit of R.

Let $f = X^n + a_{n-1}X^{n-1} + \dots + a_1X + a_0$ be a monic polynomial over R. By the *companion matrix* of f, denoted $\hat{F}$, we mean the matrix

$$\begin{bmatrix} 0 & 1 & 0 & \dots & 0 \\ 0 & 0 & 1 & \dots & 0 \\ 0 & 0 & 0 & \dots & 0 \\ \vdots & \vdots & \vdots & \dots & \vdots \\ -a_0 & -a_1 & -a_2 & \dots & -a_{n-1} \end{bmatrix}$$

It is easy to see that the characteristic polynomial of $\hat{F}$ is f.

PROPOSITION 1.2. *Let (A, b) be a single-input system over a commutative ring R and let $f = X^n + a_{n-1}X^{n-1} + \dots + a_1X + a_0$ be the characteristic polynomial of A. Then (A, b) is reachable if and only if (A, b) is pole assignable. Moreover, if (A, b) is reachable and if $h = X^n + b_{n-1}X^{n-1} + \dots + b_1X + b_0$ is any monic polynomial of degree n over R, then h is the characteristic polynomial of the matrix $A - bK$, where*

$$K = \begin{bmatrix} b_0 - a_0 & \dots & b_{n-1} - a_{n-1} \end{bmatrix} \begin{bmatrix} \hat{b} & \hat{A}\hat{b} & \dots & \hat{A}^{n-1}\hat{b} \end{bmatrix} \begin{bmatrix} b & Ab & \dots & A^{n-1}b \end{bmatrix}^{-1}$$

and where $\hat{b}$ denotes the unit column vector

$$\hat{b} = \begin{bmatrix} 0 \\ 0 \\ \vdots \\ 1 \end{bmatrix}$$

So, in the single-input case, not only is there complete control over the characteristic polynomial of a reachable system, there is even a formula for the feedback matrix necessary to obtain the desired polynomial. In the *multi-input* case, no such result exists, obviously. If it did, then all rings would be "pole-assignable" and they most definitely are not. But, for some rings, the problem can be reduced to the single-input case. It is with such rings that we shall now concern ourselves.

Call an n-dimensional system (A, B) over a ring R *feedback cyclizable* over R if there exists a vector $u \in R$ and a matrix K over R such that the single-input system $(A + BK, Bu)$ is reachable. By Proposition 2 above, if $p(X)$ is any monic polynomial of degree n over R, then there exists a matrix L over R such that the characteristic polynomial of

$$A + BK + BuL = A + B(K + uL)$$

is $p(X)$. Hence, we can arbitrarily assign the characteristic polynomial of the feedback cyclizable system (A, B). In particular, if all reachable systems over a ring R are feedback cyclizable, then all reachable systems over$R$ can, via state feedback, have their characteristic polynomials arbitrarily assigned. Heymann [7] proved that if S is a field, then all reachable systems over S are feedback cyclizable. (See [1, Lemma 3.4] for details.) It follows easily that if R is a (finite) direct sum of fields, then R has the same property. From this remark, one gets the following result of Sontag [12].

PROPOSITION 1.3. *If R is a commutative ring having only a finite number of maximal ideals, then each reachable system over R is feedback cyclizable.*

We now give a name to rings of the type mentioned in Proposition 3. Following [5], we call a ring R an FC-*ring* if and only if each reachable system over R is feedback cyclizable. (The terminology of Sontag in [12] was *suitable*). It is safe to say that most rings are not FC-rings. Indeed, Sontag himself showed [12] that if $\mathbf{R}$ is the field of real numbers and $\mathbf{R}[T]$ is the polynomial ring in one variable T over $\mathbf{R}$, then $\mathbf{R}[T]$ is not an FC-ring. It is instructive to see how that goes.

Thus, let D be a PID which is also an FC-ring. If $a, b \in D$ with $(a, b) = 1$, then the following system is reachable.

$$\left(\begin{bmatrix} 0 & 0 \\ b & 0 \end{bmatrix}, \begin{bmatrix} 1 & 0 \\ 0 & a \end{bmatrix} \right)$$

Since D is an FC-ring, there exists a matrix K over D and a vector $\mu \in D^2$ such that the determinant of the following matrix is a unit $\beta \in D$.

$$\begin{bmatrix} \mu_1 & k_{1,1}\mu_1 + k_{1,2}a\mu_2 \\ a\mu_2 & b\mu_1 + ak_{2,1}\mu_1 + a^2k_{2,2}\mu_2 \end{bmatrix}$$

where

$$\mu = \begin{bmatrix} \mu_1 \\ \mu_2 \end{bmatrix} \text{ and } K = \begin{bmatrix} k_{1,1} & k_{1,2} \\ k_{2,1} & k_{2,2} \end{bmatrix}$$

But the determinant, β, is given by

$$b\mu_1^2 + ak_{2,1}\mu_1^2 + a^2k_{2,2}\mu_1\mu_2 - ak_{1,1}\mu_1\mu_2 - a^2k_{1,2}\mu_2^2$$

which equals $b\mu_1^2 + ah$, for some element $h \in D$. Consequently, if $(a,b) = 1$, then there exist elements μ_1^2 , h , and $\beta \in D$, β a unit, such that $ah + b\mu_1^2 = \beta$.

Now, apply this to $\mathbf{R}[T]$ and to the relatively prime polynomials T and $T^2 - 1$. If $\mathbf{R}[T]$ were an FC-ring, then there must exist polynomials $h(T), \mu(T)$, and a nonzero constant $\beta \in \mathbf{R}$ such that

$$T[\mu(T)]^2 = h(T)(T^2 - 1) - \beta$$

Evaluating this at $T = \pm 1$ yields that

$$-[\mu(-1)]^2 = [\mu(+1)]^2 = -\beta \neq 0, \text{ since } \beta \in \mathbf{R}$$

As $\mu(-1)$ and $\mu(+1)$ are nonzero real numbers, their squares must be positive real numbers. The contradiction is now apparent.

Note that the last part of the argument fails in the case of the complex numbers. Now, Sontag et al [5] conjectured that if $\mathbf{C}$ is the field of complex numbers, then $\mathbf{C}[T]$ is an FC-ring. This conjecture may well be the most interesting open problem in this area. While we are unable to settle it here, we can say something.

Now an $n-$ *dimensional* system (A, B) is called *coefficient assignable* (CA) if $det(xE - (A + BF))$ runs through all monic polynomials of degree n when F runs through $R^{m\times n}$. A ring R is called a CA$_n$-*ring* when every reachable $n-$ *dimensional* system over R is coefficient assignable and a ring is called a CA-*ring* if it is a CA$_n$-ring for every integer n.

Coefficient assignability received less attention than feedback cyclization in large part because the classical approach appeared to more helpful even in the ring case. In fact, the direct search for a feedback matrix F such that the coefficients of $det(xE-(A+BF))$ become equal to prescribed values is hopeless in general. At any rate, we can now state what we know to be true about $\mathbf{C}[T]$.

THEOREM 1.4. *Let K be an algebraically closed field with $R = K[T]$, the polynomial ring in the single variable T over K. Then $K[T]$ is a CA-ring.*

The proof of this result will appear in [4].

We should remark that it has been shown in [11] that if K is a field of characteristic $p > 0$, then $K[T]$ is not an FC-ring. Consequently, if K is an algebraically closed field of characteristic $p > 0$, then $K[T]$ is a principal ideal domain which is a CA-ring, but not an FC-ring.

One essential ingredient of the proof Theorem 1.1 is that an "almost" control-canonical form exists for reachable systems over principal ideal domains. Before we give the form, we must discuss the operations of the feedback group over a ring R.

Since we seek to control the behavior of systems via state feedback, clearly an operation on the collection of all reachable systems over R of the form

$$(A, B) \longrightarrow (A + BK, B)$$

is allowed. Moreover, the behavior of the characteristic polynomial is unaltered by a change of basis in either the input space R^m or the state space R^n. These correspond to the operations

$$(A, B) \longrightarrow (A, BP^{-1}) \text{ and } (A, B) \longrightarrow (PAP^{-1}, PB)$$

where P is an invertible matrix over R. Taken together, these operations generate the feedback group acting on the set of all reachable systems over R. There is a convenient matricial interpretation of each of the basic feedback group operations. Keeping the notation of the above paragraph, regard P as being factored as a product of elementary matrices. Then, we are interested in the effect on A and B of change of basis by an elementary matrix. For a change of basis in the input space, this means adding a multiple of one column of B to another column of B. For a change of basis in the state space, this means adding a multiple of some row i of A to another row j of A, adding the same multiple of row i of B to row j of B, and subtracting the same multiple of column j of A from column i of A. The transformation $(A, B) \longrightarrow (A + BK, B)$ amounts to adding a column of B to a column of A.

We are now in a position to give the result about "almost" control-canonical forms for reachable systems over PID's.

THEOREM 1.5. *Let D be a principal ideal domain with (A, B) a reachable system over D. Then (A, B) is feedback equivalent to a system*

(A_1, B_1), where

$$A_1 = \begin{bmatrix} 0 & 1 & \cdots & \cdots & \cdots & \cdots & 0 & 0 \\ 0 & 0 & \cdots & \cdots & \cdots & \cdots 0 & 0 & \\ \vdots & \vdots & \ddots & \cdots & \cdots & \cdots & \vdots & \vdots \\ 0 & 0 & \cdots & 1 & \cdots & \cdots & 0 & 0 \\ 0 & 0 & \cdots & 0 & a & \cdots & \cdots & 0 \\ 0 & 0 & \cdots & \cdots & 0 & 1 & \cdots & 0 \\ 0 & 0 & \cdots & \cdots & \cdots & 0 & \cdots & 0 \\ \vdots & \vdots & \cdots & \cdots\cdots & \ddots & \vdots & \vdots & \\ 0 & 0 & \cdots & \cdots & \cdots & \cdots & \cdots & 1 \\ 0 & 0 & \cdots & \cdots & \cdots & \cdots & \cdots & 0 \end{bmatrix}$$

and

$$B_1 = \begin{bmatrix} 0 & \cdots & 0 & 0 & 0 \\ 0 & \cdots & 0 & 0 & 0 \\ \vdots & \ddots & \vdots & \vdots & \vdots \\ 0 & \cdots & 0 & 0 & 0 \\ 0 & \cdots & 0 & b & 0 \\ * & \cdots & * & * & 0 \\ * & \cdots & * & * & 0 \\ \vdots & \ddots & \vdots & \vdots & \vdots \\ * & \cdots & * & * & 0 \\ * & \cdots & * & * & 1 \end{bmatrix}$$

In particular, there exist elements r and s in D such that $ra + sb = 1$.

The proof is long and technical and we shall omit it. Notice that the result of Heymann follows from it. For in case D is a field since $(a, b) = 1$, at least one of a and b is nonzero. But then the matrix A_1 is easily put in the form

$$A_1 = \begin{bmatrix} 0 & 1 & 0 & \cdots & 0 & 0 & 0 \\ 0 & 0 & 1 & \cdots & 0 & 0 & 0 \\ 0 & 0 & 0 & \cdots & 0 & 0 & 0 \\ \vdots & \vdots & \vdots & \cdots & \vdots & \vdots & \vdots \\ 0 & 0 & 0 & \cdots & 0 & 1 & 0 \\ 0 & 0 & 0 & \cdots & 0 & 0 & 1 \\ 0 & 0 & 0 & \cdots & 0 & 0 & 0 \end{bmatrix}$$

and then the last column of B_1 is a *cyclic vector* for A_1.

This "almost" control-canonical form has another application to which we now turn. We have seen that$\mathbf{C}[T]$ is a CA-ring and that $\mathbf{R}[T]$ is not an FC-ring. In fact, it isn't a CA-ring [5], but that is much more difficult to prove. What *can* we say for principal ideal domains in general with regard to feedback cyclization? It is to this question that we now turn.

Here-to-fore, we have been discussing what is often called " static feedback control". This is contrasted with "dynamic feedback control", in which control of a given system is achieved by means of constructing a controller which is itself a system over the ring R . It was shown by Emre and Khargonekar that for n-dimensional systems over arbitrary commutative rings, this is *always* possible with the dimension of the constructed system having dimension at most $(n^2 - n)$. Now mathematically, this translates into the following theorem whose proof appears in the sequel .

THEOREM 1.6. *Let (A, B) be an n-dimensional system over a ring R . Then the following system is feedback cyclizable.*

$$\left(\begin{bmatrix} A & 0 \\ 0 & 0 \end{bmatrix}, \begin{bmatrix} B & 0 \\ 0 & I_{n^2-n} \end{bmatrix} \right)$$

This result has fostered research into the problem of determining the minimal dimension of the feedback system necessary to achieve feedback cyclizability. This is a very difficult problem and the interested reader is referred to [1],[2],[13],and [14] for further reading.

In this paper, we shall indicate how to easily recover the result of Emre and Khargonekar and we shall use Theorem 1.1 above to show that over PID's, it is only necessary to use a 1-dimensional dynamic controller, independent of the size of the original system.

In addition to Theorem 1.1, there is another key idea in the proof. That is the idea of associating a control process with the reachable system (A, B) and determining the number of inputs required to steer the process to all of R^n.

To be more precise, given a reachable system (A, B), consider the following control process

$$x_i = Ax_{i-1} + Bu_i$$

where the $u_i's$ are input vectors and x_i is the ith state. The following lemma was discovered by E. Kamen [8] and re-discovered by W. Schmale [11]. It says, in essence, that an n-dimensional system is feedback cyclizable if and only if the associated control process can be driven to R^n in exactly n steps. The proof is both elementary and instructive and so we include it.

LEMMA 1.7. *Let (A, B) be a reachable system over a commutative ring R. Then the system (A, B) is feedback cyclizable if and only if there exists a sequence $\mu_1, \mu_2, \ldots, \mu_n \in R^n$ such that the resulting states $x_1 = B\mu_1$, $x_2 = Ax_1 + B\mu_2$, ...,$x_n = A^{n-1}B\mu_1 + A^{n-2}B\mu_2 + \ldots + B\mu_n$ form a basis for R^n.*

Proof. ($\Rightarrow$): There exists a vector $\mu \in R^m$ and a matrix K such that $B\mu$, $(A + BK)B\mu, \ldots (A + BK)^{n-1}B\mu$ is a basis for R^n. For ease of calculation, set $\nu_1 = B\mu$, $\nu_2 = (A + BK)B\mu$, ..., $\nu_n = (A + BK)^{n-1}B\mu$.

To select the appropriate sequence of inputs, set $\mu_1 = \mu$ and for $k \geq 2$, set $\mu_k = K\nu_{k-1}$. Then one obtains the following sequence of states

$x_1 = B\mu = \nu_1$
$x_2 = BK\nu_1 + A\nu_1 = (A+BK)\nu_1 = (A+BK)B\mu = \nu_2$
$x_3 = BK\nu_2 + A(A+BK)\nu_1 = BK(A+BK)B\mu + A(A+BK)B\mu = (A+BK)^2B\mu = \nu_3$

$\vdots$

($\Leftarrow$): Conversely, suppose that we can find inputs $\mu_1, \mu_2, \ldots, \mu_n \in R^m$ such that the states $x_1 = B\mu_1, x_2 = AB\mu_1 + B\mu_2, \ldots, x_n$ form a basis for R^n. We need to find the appropriate vector μ and matrix K. Set $\mu = \mu_1$ and for $1 \leq k \leq (n-1)$, set $Kx_k = \mu_{k+1}$, and define Kx_n arbitrarily. Look at $B\mu_1$, $(A+BK)B\mu_1$, ... , $(A+BK)^{n-1}B\mu_1$. Well, one obtains the sequence $x_1 = B\mu_1$; $AB\mu_1 + BKB\mu_1 = AB\mu_1 + BKx_1 = AB\mu_1 + B\mu_2 = x_2$; $(A+BK)^2x_2 = Ax_2 + BKx_2 = Ax_2 + B\mu_3 = x_3$; ... □

How does this relate to the problem at hand? Well, let (A, B) be an n-dimensional reachable system over a ring R with associated control process $x_i = Ax_{i-1} + Bu_i$. Suppose further that exists a sequence of inputs $\mu_1, \mu_2, \ldots, \mu_{n+r}$ such that the corresponding sequence of states contains a basis $\{x_{i_1}, x_{i_2}, \ldots, x_{i_n}\}$ for R^n . For $1 \leq j \leq (n+r)$, define an input vector $\mu'_j \in R^{n+r}$ as follows:

$$\mu'_j = \begin{bmatrix} x_j \\ 0 \end{bmatrix} \text{ if } j \in \{i_1, i_2, \ldots, i_n\}$$

and

$$\mu'_j = \begin{bmatrix} x_j \\ \nu_j \end{bmatrix}$$

where as j ranges through the r integers in the set $\{1, \ldots, n+r\} \setminus \{i_1, \ldots i_n\}$, ν_j ranges through the unit vectors $\epsilon_1, \ldots, \epsilon_r$. One checks easily that the associated control process

$$x'_j = \begin{bmatrix} A & 0 \\ 0 & 0 \end{bmatrix} x'_{j-1} + \begin{bmatrix} B & 0 \\ 0 & I_r \end{bmatrix} \mu'_j \text{ , for } 1 \leq j \leq n+r$$

yields a basis $\{x'_1, x'_2, \ldots, x'_{n+r}\}$ for R^{n+r} .

Thus, by Lemma 1, for a given ring R the problem of determining the minimal dimension of the feedback system necessary to achieve feedback cyclizability becomes that of finding the least integer r, in the above notation, which works for all reachable systems over R .

For example, Theorem 1.3 can be proved using this idea.

Proof of Theorem 1.3. Let (A, B) be a reachable system with $x_i = Ax_{i-1} + B\mu_i$ the corresponding control process. If $\mu_1, \mu_2, \ldots, \mu_n$ is a sequence of n inputs, then the resulting state is

$$x_n = A^{n-1}B\mu_1 + A^{n-2}B\mu_2 + \cdots AB\mu_{n-1} + B\mu_n$$

Now, this is just a generic linear combination of the columns of the matrix

$$\begin{bmatrix} B & AB & \ldots & A^{n-1}B \end{bmatrix}$$

But, since (A, B) is a reachable system, the columns of this matrix generate R^n. In particular, the n unit basis vectors $\epsilon_1, \epsilon_2, \ldots, \epsilon_n$ are linear combinations of these columns. Putting all of this together, we have that it is possible, after a sequence of n^2 inputs, to reach a basis of R^n.

In the notation of the above discussion, the integer r was the number of vectors, *over and above* n, necessary to reach a basis of R^n. Here, we must then have that $r = n^2 - n$. This completes the proof.

We end by treating the special case of reachable systems over principal ideal domains. Here, we will show that $r = 1$ independent of the dimension of the system to be controlled. We include the proof to demonstrate the flavor and power of the methods.

THEOREM 1.8. *Let D be a principal ideal domain with (A, B) an n-dimensional reachable system over D. Then there is a sequence of $(n+1)$ inputs μ_i such that the set of states of the corresponding control process $x_i = Ax_{i-1} + B\mu_i$ contains a basis for D^n. Hence, the system*

$$\left(\begin{bmatrix} A & 0 \\ 0 & 0 \end{bmatrix}, \begin{bmatrix} B & 0 \\ 0 & 1 \end{bmatrix} \right)$$

is feedback cyclizable.

Proof. We make heavy use of Theorem 1.1. By Theorem 1.1, we can assume that our system is equivalent to one of the form

$$A_1 = \begin{bmatrix} 0 & 1 & \ldots & \ldots & \ldots & \ldots & 0 & 0 \\ 0 & 0 & \ldots & \ldots & \ldots & \ldots 0 & 0 & \\ \vdots & \vdots & \ddots & \ldots & \ldots & \ldots & \vdots & \vdots \\ 0 & 0 & \ldots & 1 & \ldots & \ldots & 0 & 0 \\ 0 & 0 & \ldots & 0 & a & \ldots & \ldots & 0 \\ 0 & 0 & \ldots & \ldots & 0 & 1 & \ldots & 0 \\ 0 & 0 & \ldots & \ldots & \ldots & 0 & \ldots & 0 \\ \vdots & \vdots & \ldots & \ldots\ldots & \ddots & \vdots & \vdots & \\ 0 & 0 & \ldots & \ldots & \ldots & \ldots & \ldots & 1 \\ 0 & 0 & \ldots & \ldots & \ldots & \ldots & \ldots & 0 \end{bmatrix}$$

and

$$B_1 = \begin{bmatrix} 0 & \dots & 0 & 0 & 0 \\ 0 & \dots & 0 & 0 & 0 \\ \vdots & \ddots & \vdots & \vdots & \vdots \\ 0 & \dots & 0 & 0 & 0 \\ 0 & \dots & 0 & b & 0 \\ * & \dots & * & * & 0 \\ * & \dots & * & * & 0 \\ \vdots & \ddots & \vdots & \vdots & \vdots \\ * & \dots & * & * & 0 \\ * & \dots & * & * & 1 \end{bmatrix}$$

where a and b are relatively prime in D and a and b are in the same row. Note that in this form, the first $m-2$ columns of the input matrix B_1 are irrelevant. So, for the proof we can assume that the system (A, B) is given by A as before, but by B as follows.

$$B = \begin{bmatrix} 0 & 0 \\ 0 & 0 \\ \vdots & \vdots \\ 0 & 0 \\ b & 0 \\ c_1 & 0 \\ c_2 & 0 \\ \vdots & \vdots \\ c_k & 0 \\ 0 & 1 \end{bmatrix}$$

By Lemma 1, we must find a sequence of inputs $\mu_1, \dots, \mu_n, \mu_{n+1} \in D^2$ such that the control process $Ax_i = x_{i-1} + B\mu_i$ contains a basis for D^n. To begin with, since $(a, b) = 1$, there exist elements r and s in D such that $ra + sb = 1$. So, set

$$\mu_1 = \begin{bmatrix} 0 \\ 1 \end{bmatrix}, \mu_2 = \begin{bmatrix} 0 \\ r \end{bmatrix}, \mu_3 = \begin{bmatrix} 0 \\ -c_1 s \end{bmatrix}, \dots,$$

and

$$\mu_{k+2} = \begin{bmatrix} 0 \\ -c_k s \end{bmatrix}, \mu_{k+3} = \begin{bmatrix} s \\ 0 \end{bmatrix}, \text{ while } \mu_i = \begin{bmatrix} 0 \\ 0 \end{bmatrix} \text{ for } i \geq k+4$$

The sequence of states $x_i = Ax_{i-1} + B\mu_i$ then becomes

$$x_1 = \begin{bmatrix} 0 \\ 0 \\ \vdots \\ \vdots \\ \vdots \\ 0 \\ 0 \\ 0 \\ 1 \end{bmatrix}, \; x_2 = \begin{bmatrix} 0 \\ 0 \\ \vdots \\ \vdots \\ \vdots \\ 0 \\ 0 \\ 1 \\ r \end{bmatrix}, \; x_3 = \begin{bmatrix} 0 \\ 0 \\ \vdots \\ \vdots \\ \vdots \\ 0 \\ 1 \\ r \\ -c_1 s \end{bmatrix}, \; \dots,$$

and

$$x_{k+1} = \begin{bmatrix} 0 \\ \vdots \\ 0 \\ 1 \\ r \\ -c_1 s \\ \vdots \\ -c_{k-1} s \end{bmatrix}, \; x_{k+2} = \begin{bmatrix} 0 \\ \vdots \\ \vdots \\ 0 \\ a \\ r \\ -c_1 s \\ \vdots \\ -c_k s \end{bmatrix},$$

and

$$x_{k+3} = \begin{bmatrix} 0 \\ \vdots \\ 0 \\ a \\ ar + bs \\ -c_1 s + c_1 s \\ \vdots \\ -c_k s + c_k s \\ 0 \end{bmatrix} = \begin{bmatrix} 0 \\ \vdots \\ 0 \\ a \\ 1 \\ 0 \\ \vdots \\ \vdots \\ 0 \end{bmatrix}, \; x_{k+4} = \begin{bmatrix} 0 \\ \vdots \\ a \\ 1 \\ 0 \\ 0 \\ \vdots \\ 0 \\ 0 \end{bmatrix}, \text{ etc.}$$

and it is apparent that the set $\{x_1, \ldots, x_{k+1}, x_{k+3}, \ldots, x_{n+1}\}$ is a basis for D^n . □

REFERENCES

1. J. Brewer, J. Bunce, and F. Van Vleck, *Linear Systems Over Commutative Ring*, Marcel Dekker, 1986.
2. J. Brewer and L. Klingler, Dynamic feedback over commutative rings, *Lin. Alg. and Its Appl.*, 98, 1988, pp. 137–168.

3. J. Brewer, L. Klingler, and W. Schmale, The dynamic feedback cyclization problem for principal ideal domains, to appear.
4. J. Brewer, L Klingler, and W. Schmale, C[y] is a CA-ring and coefficient assignment is properly weaker than feedback cyclization over a p.i.d., to appear.
5. R. Bumby, E. Sontag, H. Sussman, and W. Vasconcelos, Remarks on the pole-shifting problem over rings, *J. Pure and Appl. Alg.*, 2, 1981, pp. 113–127.
6. E. Emre and P. Khargonekar, Regulation of split linear systems over rings: Coefficient assignment and observers, *IEEE Trans. Aut. Control*, AC-27, 1982, pp. 104–113.
7. M. Heymann, Comments on pole assignment in multi-input controllable linear systems, *IEEE Trans. Aut. Control*, AC-13, 1968, pp. 748–749.
8. E. Kamen, Lectures on algebraic system theory: Linear systems over commutative rings, *NASA Contractor Report 3016*, 1978.
9. E. Kamen, *Linear systems over rings: From R. E. Kalman to the present*, Mathematical System Theory–The Influence of R. E. Kalman, Springer-Verlag, 1991.
10. W. Schmale, Feedback cyclization over certain principal ideal domains, *Int. J. Control*, 48, 1988, pp. 89–96.
11. W. Schmale, Three-dimensional feedback cyclization over C[T], *Systems and Control Letters*, 12, 1989, pp. 327–330.
12. E. Sontag, Linear systems over commutative rings: a survey, *Richerche Automat.*, 7, 1976, pp. 1–34.
13. E. Sontag, An introduction to the stabilization problem for parameterized families of linear systems, *Contemp. Math.*, 47, 1985, pp. 369–400.
14. E. Sontag and Y. Wang, Pole shifting for families of linear systems depending on at most three parameters, *Lin. Alg. and Its Appl.*, 137, 1990, pp. 3–38.

OUTPUT FEEDBACK IN DESCRIPTOR SYSTEMS *

ANGELIKA BUNSE-GERSTNER**†, VOLKER MEHRMANN**‡, AND NANCY K. NICHOLS§

Abstract. A summary is given of conditions under which a descriptor, or generalized state-space, system can be regularized by *output* feedback. Theorems are presented showing that under these conditions proportional and derivative output feedback controls can be constructed such that the closed loop system is regular and has index at most one. This property ensures the solvability of the resulting system of dynamic-algebraic equations. A canonical form is given that allows the system properties as well as the feedback to be determined. The construction procedures used to establish the theory are based only on orthogonal matrix decompositions and can therefore be implemented in a numerically stable way. A computational algorithm for improving the 'conditioning' of the regularized closed loop system is described.

1. Introduction. We consider *descriptor* or *generalized state space* systems of the form

$$E\dot{x} := Edx/dt = Ax(t) + Bu(t), \qquad x(t_0) = x_0 \tag{1.1}$$

$$y(t) = Cx(t), \tag{1.2}$$

or

$$Ex_{k+1} = Ax_k + Bu_k, \qquad x_0 \text{ given} \tag{1.3}$$

$$y_k = Cx_k, \tag{1.4}$$

where $E, A \in \mathcal{R}^{n,n}$, $B \in \mathcal{R}^{n,m}$, $C \in \mathcal{R}^{p,n}$ and $\operatorname{rank} B = m \leq n$, $\operatorname{rank} C = p \leq n$. Here $x(t)$ or $x_k \in \mathcal{R}^n$ is the state, $y(t)$ or $y_k \in \mathcal{R}^p$ is the output, and $u(t)$ or $u_k \in \mathcal{R}^m$ is the input or control of the system. In contrast to *standard* systems where $E = I$, the identity matrix, the response of a descriptor system can have a complicated structure and can even include impulsive modes.

Descriptor systems arise naturally in a variety of circumstances [21,16] and have recently been investigated in a number of papers [1,2,6,7,8,10,13,14,15,17,18,19,20,22,23,24,25,26,27,28]. The response of a descriptor system can be described in terms of the eigenstructure of the matrix pencil

$$\alpha E - \beta A. \tag{1.5}$$

---

* Partial support received from British Council/DAAD ARC Project No. 127.

† Fachbereich Mathematik und Informatik, Universität Bremen, Bibliotheksstrasse, D-2800 Bremen 33, FRG, (0421) 218-4205.

** Partial support received from SFB 343, Diskrete Strukturen in der Mathematik, Universität Bielefeld and from FSP Mathematisierung, Universität Bielefeld.

‡ Fakultät für Mathematik, Universität Bielefeld, Postfach 100131, D-4800 Bielefeld 1, FRG, (0521) 106-4798.

§ Department of Mathematics, University of Reading, Box 220, Reading, RG6 2AX, U.K., (0734) 318988.

The pencil (1.5) and the corresponding system are said to be *regular* if

$$\det(\alpha A - \beta E) \neq 0 \quad for\ some\ (\alpha, \beta) \in \mathcal{C}^2. \tag{1.6}$$

Regularity of the system guarantees the existence and uniqueness of (classical) solutions to (1.1) and (1.3) [5,27]. Many systems that are not regular can be *regularized* by proportional and/or derivative feedback. Conversely, systems that are regular can easily be transformed by linear feedback into closed loop systems that are *not* regular. It is important, therefore, to establish conditions that ensure the regularity of systems under feedback and to develop numerically reliable techniques for constructing regular closed loop systems.

The system and its pencil (1.5) are said to have *index k* if the dimension of the largest nilpotent block in the Kronecker form of $\alpha E - \beta A$ (which corresponds to an *infinite* pole) is equal to k (see [11]). A regular system has index equal to *zero* if and only if $\text{rank}(E) = n$. In this case the system is equivalent to a standard system with state and control matrices $E^{-1}A$ and $E^{-1}B$ respectively. A regular system has index less than or equal to *one* if and only if $\text{rank}[E, AS_\infty] = n$, where the columns of S_∞ span the right null space of E [13]. A regular system that has index equal to one can be transformed and separated into a purely dynamical part and a purely algebraic part. The algebraic variables can be eliminated to give a reduced order dynamical system of index zero. Such a system is thus also reducible to a standard system. Higher index descriptor systems cannot be reduced to standard systems. For such systems, impulses can arise in the response if the control is not sufficiently smooth, and the system can even lose causality [25].

In order to alter the behaviour of a descriptor system, proportional and/or derivative feedback may be used. Ideally it is desirable to select a feedback that 'regularizes' the system, that is, such that the closed loop system is regular and of index at most one. In this paper we show that such a result may be achieved by *output* feedback under certain controllability and observability conditions. We have previously derived algebraic conditions under which regularity can be achieved by proportional and/or derivative *state* feedback [3]. Here we extend these conditions to output feedback. The results are established by construction, using only unitary matrix transformations. The construction procedures can, therefore, be implemented by a numerically reliable algorithm. No assumption concerning the regularity of the original system is needed. Parts of the mathematical theory developed here have been derived concurrently by Dai [8]. Additional assumptions are required in [8], however, and the techniques used for constucting the feedback matrices in [8] are not suitable for numerical computation.

In the next section of the paper we introduce notation and give definitions of *complete* and *strong* controllability. In Section 3 we summarize the system properties that can be achieved by derivative and proportional *output* feedback under the different controllability/observability conditions. Theorems are given establishing that a system that is *completely* controllable and observable can be transformed into a *standard* system by derivative feedback. It is also established that a system that is *strongly* controllable and observable can be transformed into a regular system of index at most one (that is, a system in which impulses are excluded) by either proportional *or* derivative output feedback. Derivative feedback can be used, however, to increase the explicit degrees of freedom defining the solution space (*reachable subspace*) of the system. The construction of the required feedback matrices is obtained by transforming the system pencil to an equivalent 'canonical' form using orthogonal decompositions that are numerically stable [12]. Regularity of the open loop system is not required. Proofs of the results are omitted here. Details of the theory are presented in [4].

2. Definitions and Properties. The definitions of controllablility and observability for standard control systems can be extended to descriptor systems. Various types of controllablility/observability can be identified, however [27]. Here we investigate the properties of the generalized state-space system (1.1)-(1.2) and (1.3)-(1.4) under the following algebraic conditions:

$$\begin{aligned}
&\mathbf{C0} : \operatorname{rank}[\alpha E - \beta A, B] = n, \ \forall(\alpha,\beta) \in \mathcal{C}^2\backslash\{(0,0)\};\\
&\mathbf{C1} : \operatorname{rank}[\lambda E - A, B] = n, \ \forall\lambda \in \mathcal{C};\\
&\mathbf{C2} : \operatorname{rank}[E, AS_\infty, B] = n, \ \text{where the columns of } S_\infty \text{ span } \mathcal{N}(E).
\end{aligned} \tag{2.1}$$

For systems that are *regular*, these conditions characterize the controllability of the system. We have the following definition:

DEFINITION 2.1. *Let $\alpha E - \beta A$ be a regular pencil. Then, the triple (E, A, B) and the corresponding descriptor system are said to be* completely controllable *(C-controllable) if and only if Condition* **C0** *holds.*

We remark that a descriptor system satisfies Condition **C0**, i.e. is completely controllable, only if

$$\operatorname{rank}[E, B] = n. \tag{2.2}$$

Complete controllability ensures that for any given initial and final states $x_0, x_f \in \mathcal{R}^n$ of the system, there exists an admissible control that transfers the system from x_0 to x_f in finite time [27]. Hence, descriptor systems that are completely controllable can be expected to have similar properties to standard systems.

A weaker definition of controllability is given by the following.

DEFINITION 2.2. *Let $\alpha E - \beta A$ be a regular pencil. Then, the triple (E, A, B) and the corresponding descriptor system are said to be* strongly controllable *(S-controllable) if and only if Conditions* **C1** *and* **C2** *hold.*

We remark that C-controllability implies S-controllability. Clearly Condition **C1** follows from Condition **C0** for $\beta \neq 0$ and $\lambda = \alpha/\beta$. Condition **C2** follows from (2.2), but is weaker. In the literature, regular systems that satisfy Condition **C2** are often described as 'controllable at infinity' or 'impulse controllable' [7,13,25]. For these systems 'impulsive modes' can be excluded. A descriptor system that has a regular pencil of index less than or equal to one is always controllable at infinity, since by [13] we have $\operatorname{rank}[E, AS_\infty] = n$.

The controllability conditions are preserved under certain transformations of the system. Specifically, Conditions **C0**, **C1**, **C2**, are all preserved under non-singular 'equivalence' transformations of the pencil and under proportional state and output feedback. With the exception of Condition **C2**, these same conditions are also preserved under derivative feedback.

An example is given in [1] demonstrating that Condition **C2** is not necessarily preserved under derivative feedback. If derivative feedback is used to change the system dynamics, it is therefore necessary to be careful not to lose controllability at infinity. In the next section we investigate the use of derivative feedback to make the system regular and of index at most one. Thus, the resulting system is always controllable at infinity. *Regularity of the original system is not needed to achieve this result.*

Observability conditions for the time-invariant systems (1.1)-(1.2) and (1.3)-(1.4) can be defined as the dual of the controllability conditions. Specifically a system represented by the triple (E, A, C) is said to satisfy conditions **O0**, **O1**, **O2** if and only if the dual system, represented by the triple (E^T, A^T, C^T) satisfies the conditions **C0**, **C1**, **C2**, respectively. A regular system is defined to be *completely observable* (C–observable) if and only if Condition **O0** is satisfied and *strongly observable* (S–observable) if and only if Conditions **O1** and **O2** hold.

In the following sections we derive numerically stable techniques for constructing output feedback controllers. By duality these techniques can also be used in the construction of state estimators and observer based controllers.

3. A Canonical Form.

In this section we present our first main theorem, which gives a canonical form for systems (1.1)–(1.2) or (1.3)–(1.4) that can be computed using only unitary equivalence transformations. This canonical form is used to establish the main results in Section 4.

THEOREM 3.1. *If $E, A \in \mathcal{C}^{n,n}$, $B \in \mathcal{C}^{m,n}$, $C \in \mathcal{C}^{p,n}$ then there exist unitary matrices $U, V \in \mathcal{C}^{n,n}$, $W \in \mathcal{C}^{m,m}$, $Y \in \mathcal{C}^{p,p}$ such that*

$$U^H E V = \begin{bmatrix} \Sigma_E & 0 \\ 0 & 0 \end{bmatrix} \begin{matrix} t_1 \\ n - t_1 \end{matrix}, \quad U^H B W = \begin{bmatrix} B_{11} & B_{12} & 0 \\ \hat{B}_{21} & 0 & 0 \end{bmatrix} \begin{matrix} t_1 \\ n - t_1 \end{matrix},$$

$$Y^H C V = \begin{bmatrix} C_{11} & \hat{C}_{12} \\ C_{21} & 0 \\ 0 & 0 \end{bmatrix} \begin{matrix} \ell_1 \\ \ell_2 \\ \ell_3 \end{matrix}, \quad U^H A V = \begin{bmatrix} A_{11} & \hat{A}_{12} \\ \hat{A}_{21} & \hat{A}_{22} \end{bmatrix} \begin{matrix} t_1 \\ n - t_1 \end{matrix}, \tag{3.1}$$

where

$$\hat{A}_{22} = \left[\begin{array}{ccccc} A_{22} & A_{23} & A_{24} & 0 & 0 \\ A_{32} & A_{33} & A_{34} & \Sigma_{35} & 0 \\ A_{42} & A_{43} & \Sigma_{44} & 0 & 0 \\ 0 & \Sigma_{53} & 0 & 0 & 0 \\ 0 & 0 & 0 & 0 & 0 \end{array} \right] \begin{array}{c} t_2 \\ t_3 \\ t_4 \\ t_5 \\ t_6 \end{array}, \quad \hat{B}_{21} = \left[\begin{array}{c} B_{21} \\ B_{31} \\ 0 \\ 0 \\ 0 \end{array} \right] \begin{array}{c} t_2 \\ t_3 \\ t_4 \\ t_5 \\ t_6 \end{array},$$

$$\hat{C}_{12} = \left[\begin{array}{ccccc} C_{12} & C_{13} & 0 & 0 & 0 \end{array} \right] \ell_1 ,$$

$$\hat{A}_{21} = \left[\begin{array}{c} A_{21} \\ A_{31} \\ A_{41} \\ A_{51} \\ A_{61} \end{array} \right] \begin{array}{c} t_2 \\ t_3 \\ t_4 \\ t_5 \\ t_6 \end{array}, \hat{A}_{12} = \left[\begin{array}{ccccc} A_{12} & A_{13} & A_{14} & A_{15} & A_{16} \end{array} \right] t_1 . \tag{3.2}$$

Here A_{11}, Σ_E are $t_1 \times t_1$ matrices, the row dimensions of the matrices are as depicted, and the column dimensions are $[t_1, s_2, t_5, t_4, t_3, s_6]$ for $U^H EV$, $U^H AV$, $Y^H CV$ and $[k_1, k_2, k_3]$ for $U^H BW$.

The matrices $\Sigma_E, \Sigma_{35}, \Sigma_{44}, \Sigma_{53}$ are non-singular diagonal matrices, B_{12} has full column rank, C_{21} has full row rank and the matrices

$$\left[\begin{array}{c} B_{21} \\ B_{31} \end{array} \right] \in \mathcal{C}^{k_1,k_1}, \ \left[\begin{array}{c} C_{12}, C_{13} \end{array} \right] \in \mathcal{C}^{\ell_1,\ell_1}$$

with $k_1 = t_2 + t_3$ and $\ell_1 = s_2 + t_5$ are non-singular.

Proof. The proof is by construction via Algorithm 1 given in [4]. □

In theory we assume that B, C have full rank. We assume, therefore, that in the canonical form (3.1) we have $k_3 = \ell_3 = 0$, i.e. the column of zero blocks in $U^H BW$ and the row of zero blocks in $Y^H CV$ are void.

From Theorem 3, it follows that the system (1.1)–(1.2) or (1.3)–(1.4) can be made regular and of index at most one if and only if the last block rows of $\hat{A}_{21}, \hat{A}_{22}$ and the last block columns of $\hat{A}_{12}, \hat{A}_{22}$ are void, that is, if and only if $s_6 = t_6 = 0$. This must be the case if the Conditions **C2** and **O2** hold. This result provides the basis for the main theorems, which are presented in the next section.

4. Derivative and Proportional Feedback for Descriptor Systems. In this section we discuss conditions under which we can alter the structure of the system pencil (1.5) by the use of derivative and/or proportional output feedback. We show that if (E, A, B, C) satisfies Condition **C0** and Condition **O0**, then the system (1.1)-(1.2) or (1.3)-(1.4) can be transformed into a *completely controllable, completely observable standard* system by derivative feedback. We show also that if a system satisfies Conditions **C1**, **C2**, **O1** and **O2**, then a closed loop system that is *regular, strongly controllable, strongly observable, and of index at most one* can be obtained by derivative or proportional feedback. With derivative feedback, however, the explicit degrees of freedom describing the reachable subspace of the system (corresponding to the number of finite poles of the closed loop system) can be increased to a maximum.

In the main theorem we establish conditions that guarantee that there exist matrices F and G such that the matrix pencil $\alpha(E+BGC)-\beta(A+BFC)$ is regular and of index at most one.

THEOREM 4.1. *Let $E, A \in \mathcal{R}^{n,n}$, $B \in \mathcal{R}^{n,m}$, $C \in \mathcal{R}^{p,n}$ with* $\operatorname{rank}(B) = m \leq n$ *and* $\operatorname{rank}(C) = p \leq n$. *If Conditions* **C2** *and* **O2** *hold, then there exist matrices $F, G \in \mathcal{R}^{m,p}$ such that the matrix pencil $\alpha(E+BGC)-\beta(A+BFC)$ is regular, has index at most one and* $\operatorname{rank}(E+BGC) = r$, *where* $\operatorname{rank}(E) \leq r \leq q \equiv \operatorname{rank}(E) + t_2$. *(Here t_2 is defined by the decomposition of Theorem 3.)*

Proof. The proof follows from Theorem 3 by construction, using singular value matrix decompositions [4]. □

Theorem 4.1 provides the key step in the proofs of the following theorems. The main results can be stated as follows.

THEOREM 4.2. *There exists a real feedback control $u = -G\dot{x} + v$ or $u_k = -Gx_{k+1} + v_k$ such that the continuous or discrete closed loop system defined by $(E+BGC, A, B, C)$ is C-controllable, C-observable, and the matrix $E+BGC$ is non-singular if and only if the system (E, A, B, C) satisfies Conditions* **C0** *and* **O0**.

We remark that if the system matrix $S = E+BGC$ is non-singular, then the corresponding closed loop system is equivalent to a standard system. Transformation to this standard form may not be numerically reliable, however, if $S = E+BGC$ is ill-conditioned with respect to inversion. In a later section of this paper we discuss techniques for selecting G to provide an optimally conditioned 'regularization' of the system.

THEOREM 4.3. *There exists a real feedback control $u = -G\dot{x} + v$ or $u_k = -Gx_{k+1} + v_k$ such that the continuous or discrete closed loop system defined by $(E+BGC, A, B, C)$ is S–controllable, S-observable and the system pencil $\alpha(E+BGC) - \beta A$ is regular, has index at most one and* $\operatorname{rank}(E+BGC) = q \equiv \operatorname{rank}(E) + t_2$ *if the system (E, A, B, C) satisfies Conditions* **C1 C2 O1** *and* **O2**.

From Theorem 4.3 we conclude that systems that satisfy Conditions **C1 C2 O1** and **O2** can be transformed by derivative feedback into completely controllable, completely observable, reduced order, standard systems with maximal dimension.

THEOREM 4.4. *There exists a real feedback control $u = Fx + v$ or $u_k = Fx_k + v_k$ such that the continuous or discrete closed loop system defined by $(E, A+BFC, B, C)$ is S-controllable and S-observable and the system pencil $\alpha E - \beta(A+BFC)$ is regular and of index at most one if and only if the system (E, A, B, C) satisfies Conditions* **C1 C2 O1** *and* **O2**.

From Theorem 4.4 we see that systems that satisfy Conditions **C1 C2 O1** and **O2** can also be transformed by proportional output feedback into regular systems of index at most one and hence into completely controllable, completely observable, reduced order *standard* systems. The order of the standard system is minimal, however, being equal to rank(E). The degrees of freedom in the reduced order dynamical system thus do not reflect the dimension of the reachable subspace of the original descriptor system.

In the next section of the paper we discuss techniques for 'regularizing' the descriptor system so as to ensure that the dynamic part of the closed loop system is as well-conditioned as possible.

5. Improved Conditioning. In previous sections of this paper we have examined conditions under which the descriptor systems (1.1)-(1.2) and (1.3)-(1.4) can be 'regularized' by derivative and proportional feedback, that is, conditions that ensure that a closed loop system can be constructed that is regular and of index at most one, and is S–controllable and S–observable. Regularity of the original system is not required, and the construction procedures are based on numerically stable techniques.

In practice it is desirable not only that the closed loop descriptor system is regular, but also that it is 'well-conditioned' in the sense that the reduction to standard form is computationally reliable. In this section of the paper we describe a computational technique for generating feedback matrices F and G such that the closed loop system is optimally 'conditioned'.

It is our aim to choose F, G such that the closed loop pencil

$$\alpha(E + BGC) - \beta(A + BFC)$$

is regular and of index at most one and is unitarily equivalent to

$$\alpha \begin{bmatrix} \mathcal{E}_{11} & 0 \\ 0 & 0 \end{bmatrix} - \beta \begin{bmatrix} \mathcal{A}_{11} & \mathcal{A}_{12} \\ \mathcal{A}_{21} & \mathcal{A}_{22} \end{bmatrix} \tag{5.1}$$

where $\mathcal{E}_{11} \in \mathcal{R}^{t_1+s,t_1+s}$, $\mathcal{A}_{22} \in \mathcal{R}^{n-t_1-s,n-t_1-s}$ are as well-conditioned with respect to inversion as possible. (A matrix M is well-conditioned with respect to inversion if its condition number cond$(M) := ||M||\, ||M^{-1}||$ is small. Here and in the following, $||\ ||$ is the spectral norm.) If the conditioning of these matrices is improved, then the closed loop system is less sensitive to perturbations and also has better performance characteristics (see for example [20]).

Using the canonical form of Theorem 3 it can be shown that with an appropriate choice of the feedback matrices F and G, the transformed pencil (5.1) can be attained where

$$\mathcal{A}_{22} = \begin{bmatrix} A_{33} + \hat{F} & A_{34} \\ A_{43} & A_{44} \end{bmatrix}. \tag{5.2}$$

and $\mathcal{E}_{11}$ is unitarily equivalent to

$$\tilde{\mathcal{E}}_{11} = \begin{bmatrix} E_{11} + \hat{G} & E_{12} \\ E_{21} & E_{22} \end{bmatrix}. \tag{5.3}$$

The required matrices F and G are determined from the matrices $\hat{F}$ and $\hat{G}$, and these must be selected such that $\mathcal{A}_{22}$ and $\tilde{\mathcal{E}}_{11}$ are as well-conditioned as possible.

The optimal choice of both feedback matrices F and G can, therefore, be determined if we can solve the following problem.

PROBLEM 1. *For a given matrix* $M = \begin{bmatrix} M_{11} & M_{12} \\ M_{21} & M_{22} \end{bmatrix}$ *determine* Δ *of the same size as* M_{11} *such that* $\operatorname{cond}\left(\begin{bmatrix} M_{11}+\Delta & M_{12} \\ M_{21} & M_{22} \end{bmatrix}\right)$ *is minimal.*

In general this is an open problem. Partial solutions that give the optimum if $\begin{bmatrix} M_{21} & M_{22} \end{bmatrix}$ is void, are derived in [9]. In the following we compute the optimizing feedbacks for an upper bound on the condition number. To do this we determine unitary matrices U_M, V_M such that

$$\hat{M} = U_M^H \begin{bmatrix} M_{11}+\Delta & M_{12} \\ M_{21} & M_{22} \end{bmatrix} V_M = \begin{bmatrix} \Delta_{11} & \Delta_{12} & \hat{M}_{13} & 0 \\ \Delta_{21} & \Delta_{22} & \hat{M}_{23} & \Sigma_{24} \\ \hat{M}_{31} & \hat{M}_{32} & \Sigma_{33} & 0 \\ 0 & \Sigma_{42} & 0 & 0 \end{bmatrix}, \tag{5.4}$$

with arbitrary blocks $\Delta_{11}, \Delta_{12}, \Delta_{21}, \Delta_{22}$ and non-singular diagonal matrices $\Sigma_{24}, \Sigma_{42}, \Sigma_{33}$. We then write $\hat{M} = RDL$, from which we obtain

$$\operatorname{cond}(\hat{M}) \leq \operatorname{cond}(L)\operatorname{cond}(D)\operatorname{cond}(R),$$

where

$$D = \begin{bmatrix} \tilde{\Delta}_{11} & \Delta_{12} & 0 & 0 \\ \Delta_{21} & \Delta_{22} & 0 & \Sigma_{24} \\ 0 & 0 & \Sigma_{33} & 0 \\ 0 & \Sigma_{42} & 0 & 0 \end{bmatrix}, \tag{5.5}$$

$\tilde{\Delta}_{11} := \Delta_{11} - \hat{M}_{13}\Sigma_{33}^{-1}\hat{M}_{31}$, and $\operatorname{cond}(L)$ and $\operatorname{cond}(R)$ are independent of

$$\tilde{\Delta} = \begin{bmatrix} \tilde{\Delta}_{11} & \Delta_{12} \\ \Delta_{21} & \Delta_{22} \end{bmatrix}.$$

Since we have to invert the diagonal matrices Σ_{33}, Σ_{24} and Σ_{42} to obtain D^{-1} and since $\tilde{\Delta}$ is free, we choose $\Delta_{12}, \Delta_{21}, \Delta_{22}$ to be zero and $\tilde{\Delta}_{11} = \alpha I$, where α is the minimum singular value of $\Sigma = \begin{bmatrix} \Sigma_{24} & 0 & 0 \\ 0 & \Sigma_{42} & 0 \\ 0 & 0 & \Sigma_{33} \end{bmatrix}$.

We then have the feedback

$$\Delta = \begin{bmatrix} \Delta_{11} & \Delta_{12} \\ \Delta_{21} & \Delta_{22} \end{bmatrix} = \begin{bmatrix} \alpha I + \hat{M}_{13}\Sigma_{33}^{-1}\hat{M}_{31} & 0 \\ 0 & 0 \end{bmatrix}. \tag{5.6}$$

It is possible that this choice does not improve the condition number; in this case we choose Δ to be the zero matrix.

Performing this method for $\mathcal{E}_{11}, \mathcal{A}_{22}$ we can improve the conditioning of the system. An Algorithm is given in [4].

6. Conclusion. We investigate here the use of proportional and derivative *output* feedback for regularizing a descriptor, or generalized state-space system. We define various algebraic conditions guaranteeing that a descriptor system can be transformed by output feedback into a closed loop system that is regular and of index at most one.

Conditions are given under which the system can be transformed into a completely controllable and completely observable *standard* system (of full dimension) by a combination of derivative and proportional output feedback.

Conditions are also given under which the system can be transformed by proportional and derivative output feedback into a *strongly* controllable and *strongly* observable system that is regular and of index at most one with precisely r *finite* poles, where r lies in a region of integers determined from E, A, B, C via a canonical form. Such a system is 'impulse controllable' and can be transformed into a reduced-order controllable and observable *standard* system of precise dimension r.

The proofs of these results do not require regularity of the original open loop system. Furthermore, the procedure for constructing the feedback matrices that regularize the closed loop system are based on unitary matrix decompositions and are numerically stable. In practice it is desirable not only that the closed loop descriptor system is regular, but also 'well-conditioned' in the sense that the reduction to standard form is computationally reliable. We show here that the feedback matrices that regularize the system can also be chosen to improve the 'conditioning' of the closed loop system. A computational algorithm for achieving these results is presented in [4], where detailed proofs of the theorems are also given.

REFERENCES

[1] A. Bunse-Gerstner, V. Mehrmann, N.K. Nichols, "Derivative Feedback for Descriptor Systems", FSP Mathematisierung, Universität Bielefeld, Bielefeld, F.R.G., *Materialien LVIII*,1989, presented at the IFAC International Workshop on System Structure and Control: State-space and Polynomial Methods, Prague, September 1989.

[2] A. Bunse-Gerstner, V. Mehrmann, N.K. Nichols, "On Derivative and Proportional Feedback Design for Descriptor Systems", in *Signal Processing, Scattering and Operator Theory, and Numerical Methods*, Proceedings of the Intern. Symposium MTNS-89, (M.A. Kaashoek et al edtrs.), Vol.III, Birkhäuser, Basel, 1990, pp.437-446

[3] A. Bunse-Gerstner, V. Mehrmann, N.K. Nichols, "Regularization of Descriptor Systems by Derivative and Proportional State Feedback," *SIAM J. Matrix Analysis Appl.*, Vol.13, 1992, pp. 46-67.

[4] A. Bunse-Gerstner, V. Mehrmann, N.K. Nichols, "Numerical Methods for the Regularization of Descriptor Systems by Output Feedback", Institute for Mathematics and its Applications, University of Minnesota, IMA Preprint Series No. 987, July, 1992.

[5] S.L. Campbell," Singular Systems of Differential Equations " *Pitman, San Francisco*, 1980.

[6] M. Christodoulou," Decoupling in the Design and Synthesis of Singular Systems" *Automatica*, Vol. 22, No. 2 , 1986, pp. 245–249.

[7] D.J. Cobb," Feedback and Pole Placement in Descriptor Variable Systems" *Int. J. Control*, Vol. 33, No. 6, 1981, pp. 1135–1146.

[8] L.Dai," Singular Control Systems", Lecture Notes in Control and Information Sciences, Vol. 118, Springer Verlag, Berlin 1989

[9] L. Elsner, C. He, V. Mehrmann, " Minimizing the Norm, the Norm of the Inverse and the Condition Number of a Matrix by Completion", Preprint 92-028, Sonderforschungsbereich 343, 'Diskrete Strukturen in der Mathematik', Universität Bielefeld, to appear in Jour. Num. Lin. Algebra with Appl.

[10] L.R. Fletcher, J. Kautsky, N.K. Nichols," Eigenstructure Assignment in Descriptor Systems" *IEEE Trans. Autom. Control*, AC–31, 1986, pp. 1138–1141.

[11] F.R. Gantmacher," Theory of Matrices," Vol. I, II *Chelsea, New York*, 1959.

[12] G.H. Golub, C.F. Van Loan ," Matrix Computations" *North Oxford Academic, Oxford*, 1983.

[13] J. Kautsky, N.K. Nichols, E.K.-W. Chu ," Robust Pole Assignment in Singular Control Sytems" *Lin. Alg. Appl.*, Vol. 121, 1989, pp. 9-37.

[14] V. Kucera, D. Zagalak, "Fundamental Theorem of State Feedback for Singular Systems", *Automatica*, Vol. 24, 1988, pp. 653–658

[15] F.L. Lewis," Descriptor Systems: Fundamental Matrix, Reachability and Observability Matrices, Subspaces" *Proceedings of 23rd Conference on Decision and Control, Las Vegas, NV*, December 1984, pp. 293–298.

[16] D.G. Luenberger, "Dynamic Equations in Descriptor Form" *IEEE Trans. Autom. Control*, Vol. AC–22, 1977, pp. 312–321.

[17] V. Mehrmann," Existence, Uniqueness and Stability of Solutions to Singular Linear Quadratic Control Problems " *Lin. Alg. Appl.*, Vol. 121, 1989, 291-331

[18] R. Mukundan, W. Dayawansa," Feedback Control of Singular Systems – Proportional and Derivative Feedback of the State" *Int. J. Syst. Sci.*, Vol. 14, 1983, pp. 615–632.

[19] K. Ozcaldiran, " Geometric Notes on Descriptor Systems", Proceedings of the 27th, CDC, Los Angeles, Ca, 1987

[20] D.W. Pearson, M.J. Chapman, D.N. Shields," Partial Singular Value Assignment in the Design of Robust Observers for Discrete Time Descriptor Systems" *IMA. J. of Math. Control and Inf.*, Vol. 5, 1988, pp. 203–213.

[21] H.H. Rosenbrock," Structural Properties of Linear Dynamic Systems" *Int. J. Control*, Vol. 20, No. 2, 1974,pp. 191–202.

[22] M.A. Shayman," Pole Placement by Dynamic Compensation for Descriptor Systems" *Automatica*, Vol. 24, 1988, pp. 279–282.

[23] M.A. Shayman, Z. Zhou," Feedback Control and Classification of Generalized Linear Systems" *IEEE Trans. Autom. Control*, Vol. AC–32, No. 6, 1987, pp. 483–494.

[24] P. Van Dooren,"The Generalized Eigenstructure Problem in Linear System Theory," *IEEE Transactions on Automatic Control*, Vol. 6, 1981, pp. 111-129.

[25] G.C. Verghese, B.C. Lévy, T. Kailath ," A General State Space for Singular Systems" *IEEE Trans. Autom. Control*, AC–26, No. 4, 1981, pp. 811–831.

[26] G.C. Verghese, P. Van Dooren, T. Kailath," Properties of the System Matrix of a Generalized State Space System" *Int. J. Control*, Vol. 30, 1979, pp. 235–243.

[27] E.L. Yip, R.F. Sincovec, " Solvability, Controllability and Observability of Continuous Descriptor Systems" *IEEE Trans. Autom. Control*, AC–26, 1981, pp. 702–707.

[28] Z. Zhou, M.A. Shayman, T.-J. Tarn, " Singular Systems: A New Approach in the Time Domain" *IEEE Trans. Autom. Control*, Vol. AC–32, No. 1, 1987, pp. 42–50.

ON REALIZATION THEORY FOR GENERALIZED STATE-SPACE SYSTEMS OVER A COMMUTATIVE RING

J. DANIEL COBB*

Abstract. The problem of finding a state-space realization for a given rational matrix over a commutative ring is considered. To simplify the problem, we assume a certain factored structure for the denominator polynomials in the matrix. Our main results state that this class of matrices is a module which can be decomposed into two independent and isomorphic submodules, each realizable via existing results for strictly proper matrices. Any rational matrix with factored denominators can be realized through this decomposition.

Key words. algebraic systems, realization theory, singular systems, singular perturbation

1. Introduction. The theory of generalized state-space systems has been developed extensively over the past decade [1]. Most of this work has centered around dynamic system equations of the form

$$\begin{aligned} E\dot{x} &= Ax + Bu \\ y &= Cx \end{aligned} \tag{1}$$

and their transfer function matrices

$$H(s) = C(sE - A)^{-1}B, \tag{2}$$

where E, A, B, and C are real matrices, E and A square. In this paper we initiate the study of such systems over a commutative ring R. In particular, we are interested here in the algebraic aspects of the problem of finding a dynamic system (1) which realizes a given rational matrix (2) over R.

One motivation for examining generalized state-space systems in an algebraic setting comes from singular perturbation theory. If the class of rings R to be considered is chosen to reflect a "perturbational" structure in (1), then those classical results in singular perturbation theory which rest on purely algebraic arguments can be exposed. Several examples of "perturbational" rings follow:

1) Convergent sequences $\{x_k\}$ in $\mathbb{R}$ under the equivalence

$$\{x_k\} \approx \{y_k\} \text{ iff } x_k = y_k \text{ for large } k.$$

* Department of Electrical and Computer Engineering, University of Wisconsin-Madison, 1415 Johnson Drive, Madison, Wisconsin 53706–1691. This work was supported by NSF grant #ECS-8920081.

Addition and multiplication are defined by

$$[x_k] + [y_k] = [x_k + y_k]$$

$$[x_k][y_k] = [x_k y_k].$$

2) C^r functions $x : U_x \to \mathbb{R}$, where $U_x \subset \mathbb{R}^k$ is a neighborhood of 0, under the equivalence

$$x \approx y \text{ iff } \exists \text{ a neighborhood } U \subset U_x \cap U_y \text{ of } 0 \text{ s.t. } x \equiv y \text{ on } U.$$

$$[x] + [y] = [x + y]$$

$$[x][y] = [xy]$$

3) Analytic functions $x : U_x \to \mathbb{C}$ with $U_x \subset \mathbb{C}$ connected, using the same equivalence as in 2). In this case, any two equivalent functions are restrictions of a single function.

An important feature shared by these examples is that each ring contains an ideal P satisfying the property that $1 + x$ is a unit of R whenever $x \in P$. In example 1), one choice of P is the family of sequences converging to 0. In 2) and 3), the set of all x with $x(0) = 0$ plays the same role. We need not restrict ourselves to these examples, however, to find such ideals. Indeed, for any ring R, the Jacobson radical J is characterized by

$$J = \{x \in R | 1 + xy \text{ is a unit for every } y \in R\}.$$

The ideals mentioned relative to 1)-3) are just the Jacobson radicals for those rings.

In order to talk about rational matrices over R, we need to first consider the ring $R[s]$ of polynomials over R and the subset $D \subset R[s]$ given by

$$D = \{\Delta \in R[s] | \Delta(s) = u(s^r + a_{r-1}s^{r-1} + \cdots + a_0)(b_q s^q + \cdots + b_1 s$$
(3)
$$+1), b_i \in J, u \text{ is a unit of } R\}$$

D may be viewed as the product of the two sets

$$D_s = \{\Delta \in R[s] | \Delta(s) = u(s^r + a_{r-1}s^{r-1} + \cdots + a_0), u \text{ a unit}\}$$

and

$$D_f = \{\Delta \in R[s] | \Delta(s) = u(b_q s^q + \cdots + b_1 s + 1), b_i \in J, u \text{ a unit}\}.$$

Note that D is multiplicatively closed. We now established another useful property.

THEOREM 1.1. *D contains no zero divisor of $R[s]$.*

Proof. It is easy to see that each $\Delta \in D$ is of the form

$$(4) \quad \Delta(s) = v(c_{q+r}s^{q+r} + \cdots + c_{r+1}s^{r+1} + s^r + c_{r-1}s^{r-1} + \cdots + c_0),$$

where $c_{r+1}, \ldots, c_{q+r} \in J$ and v is a unit. Suppose $\lambda(s) = d_k s^k + \cdots + d_0$ is any polynomial over R such that $\Delta\lambda = 0$. Then

$$\Gamma \begin{bmatrix} d_0 \\ \vdots \\ d_k \end{bmatrix} = 0,$$

where

$$\Gamma = \begin{bmatrix} 1 & c_{r-1} & \cdots & c_{r-k} \\ c_{r+1} & \ddots & \ddots & \vdots \\ \vdots & \ddots & \ddots & c_{r-1} \\ c_{r+k} & \cdots & c_{r+1} & 1 \end{bmatrix}.$$

(Whenever $i < 0$ or $i > q+k$, c_i is replaced by 0.) A simple calculation shows that $\det \Gamma$ is a unit; hence, $\lambda = 0$. □

Thus the R-module of fractions

$$D^{-1}R[s] = \left\{ \frac{x}{\Delta} | x \in R[s], \Delta \in D \right\}$$

is formed by imposing the equivalence relation

$$(x_1, \Delta_1) \approx (x_2, \Delta_2) \text{ iff } x_1\Delta_2 = x_2\Delta_1$$

on $R[s] \times D$ and passing to the quotient set, using the standard operations for rational functions. (See [2, p. 36].) Note that $H(s)$ in (2) belongs to $D^{-1}R[s]$ whenever $\det(sE - A) \in D$.

2. Decomposition of $D^{-1}R[s]$. The main result of this section shows that the R-module $D^{-1}R[s]$ has a natural decomposition commensurate with the factored polynomial form assumed in the definition (4) of D. Let

$$\mathcal{H}_s = \{h \in D^{-1}R[s] | h \text{ is strictly proper and has a representative with denominator in } D_s\}$$

and

$$\mathcal{H}_f = \{h \in D^{-1}R[s] | h \text{ has a representative with denominator in } D_f\}.$$

Clearly, $\mathcal{H}_s$ and $\mathcal{H}_f$ are R-submodules of $D^{-1}R[s]$, and $R[s]$ is an R-submodule of $\mathcal{H}_f$

THEOREM 2.1. $D^{-1}R[s] = \mathcal{H}_s \oplus \mathcal{H}_f$

Proof. Suppose

$$h(s) = \frac{c_k s^k + \cdots + c_0}{u(s^r + a_{r-1}s^{r-1} + \cdots + a_0)(b_q s^q + \cdots + b_1 s + 1)},$$

where $k \geq \max\{q+r-1, r\}$, $b_i \in J$, and u is a unit. (There is no loss of generality in constraining k, since some of the leading c_i's may be 0.) Let

$$\begin{bmatrix} W \\ Y \end{bmatrix} = \begin{bmatrix} 1 & 0 & \cdots & 0 \\ b_1 & \ddots & \ddots & \vdots \\ \vdots & & \ddots & 1 \\ b_1 & & & b_1 \\ 0 & & \ddots & \vdots \\ \cdot & & \ddots & b_q \\ \cdot & & & 0 \\ & & & \vdots \\ \cdot & & & \\ 0 & & \cdots & 0 \end{bmatrix}, \quad \begin{bmatrix} X \\ Z \end{bmatrix} = \begin{bmatrix} a_0 & 0 & \cdots & 0 \\ \vdots & \ddots & \ddots & \vdots \\ a_{r-1} & & \cdot & 0 \\ 1 & \cdot & & a_0 \\ 0 & \cdot & \cdot & \vdots \\ \vdots & \ddots & \cdot & a_{r-1} \\ 0 & \cdots & 0 & 1 \end{bmatrix},$$

where $W \in R^{r\times r}, X \in R^{r\times(k-r+1)}, Y \in R^{(k-r+1)\times r}, Z \in R^{(k-r+1)\times(k-r+1)}$. Note that $\det W = \det Z = 1$, so

$$\det \begin{bmatrix} W & X \\ Y & Z \end{bmatrix} = \det(I - YW^{-1}XZ^{-1}). \tag{5}$$

Since $Y \in J^{(k-r+1)\times r}$, the determinant (5) is a unit. Let

$$\begin{bmatrix} d_0 \\ \vdots \\ d_{r-1} \\ e_0 \\ \vdots \\ e_{k-r} \end{bmatrix} = \begin{bmatrix} W & X \\ Y & Z \end{bmatrix}^{-1} \begin{bmatrix} c_0 \\ \vdots \\ c_k \end{bmatrix}. \tag{6}$$

Equation (6) is equivalent to writing $h = h_s + h_f$, where

$$h_s(s) = \frac{d_{r-1}s^{r-1} + \cdots + d_0}{s^r + a_{r-1}s^{r-1} + \cdots + a_0} \tag{7}$$

and

$$h_f(s) = \frac{e_{k-r}s^{k-r} + \cdots + e_0}{b_q s^q + \cdots + b_1 s + 1}. \tag{8}$$

Clearly, $h_s \in \mathcal{H}_s$ and $h_f \in \mathcal{H}_f$.

To verify independence of $\mathcal{H}_s$ and $\mathcal{H}_f$, choose $h \in \mathcal{H}_s \cap \mathcal{H}_f$. Then h has both forms (7) and (8) simultaneously. Thus $q = r = 0$ and, since h is strictly proper, $h = 0$. $\square$

3. State-Space Realization in $D^{-1}R[s]$. Given $H \in (D^{-1}R[s])^{p\times m}$, a 4-tuple (E, A, B, C) of matrices over R is said to be a realization of H if $H(s) = C(sE - A)^{-1}B$. For $E \in \mathcal{H}_s^{p\times m}$, it is shown in [3, Ch. 4] that there exists a realization of the form (I, A, B, C). If $H \in \mathcal{H}_f^{p\times m}$, we may exploit this result with the aid of the following construction. Let $d : (D^{-1}R[s])^{p\times m} \to (D^{-1}R[s])^{p\times m}$ be defined by applying the map $h(s) \to -\frac{1}{s}h(\frac{1}{s})$ to each entry of the matrix. (The manipulation of the indeterminate s is formal.)

THEOREM 3.1.

1) d is a module automorphism on $(D^{-1}R[s])^{p\times m}$.

2) d maps $\mathcal{H}_f^{p\times m}$ isomorphically onto $\mathcal{H}_s^{p\times m}$.

Proof. 1) Direct calculation shows that d is linear and that $d \circ d$ is the identity map.

2) It suffices to show that $d(\mathcal{H}_f^{p\times m}) \subset \mathcal{H}_s^{p\times m}$. Let $h \in \mathcal{H}_f$. Using the notation in (8), and assuming $k \geq q + r$, we obtain

$$-\frac{1}{s}h\left(\frac{1}{s}\right) = \frac{-e_0 s^{k-r} - \cdots - e_{k-r}}{s^{k+1} + b_1 s^k + \cdots + b_q s^{k-r-q+1}},$$

which belongs to $\mathcal{H}_s$. □

Suppose $H \in \mathcal{H}_f^{p\times m}$. Then $d(H)$ has a realization (I, A, B, C). In fact, this realization satisfies

$$C(sA - I)^{-1}B = -\frac{1}{s}C\left(\frac{1}{s}I - A\right)^{-1}B = -\frac{1}{s}d(H)\left(\frac{1}{s}\right) = H(s),$$

so (A, I, B, C) is a realization of H. Combining Theorems 2.1 and 3.1 enables us to realize any $H \in (D^{-1}R[s])^{p\times m}$. Indeed, we may decompose $H = H_s + H_f$ and choose realizations (I, A_s, B_s, C_s) and (I, A_f, B_f, C_f) of H_s and $d(H_f)$, respectively. Defining

$$E = \begin{bmatrix} I & 0 \\ 0 & A_f \end{bmatrix}, \; A = \begin{bmatrix} A_s & 0 \\ 0 & I \end{bmatrix}, \; B = \begin{bmatrix} B_s \\ B_f \end{bmatrix}, \; C = [C_s \;\; C_f],$$

we have

$$C(sE - A)^{-1}B = C_s(sI - A_s)^{-1}B_s + C_f(sA_f - I)^{-1}B_f = H_s + H_f = H,$$

so (E, A, B, C) is a realization of H.

4. Conclusions. We have presented a general algebraic treatment of the state-space realization problem for rational matrices H over a commutative ring, provided that the denominators of the entries of H are of the form in (4). For arbitrary rational matrices a factorization theory would have to be developed for polynomials whose leading coefficients are members of the Jacobson radical, followed by a unit.

Another important problem that shows promise is that of developing an algebraic version of the Weierstrass decomposition for regular matrix pencils over a ring. Here the definition of regularity would undoubtedly be critical. One possibility would be to call a pencil regular if its determinant polynomial is of the form (4).

REFERENCES

[1] F.L. Lewis, *A survey of linear singular systems*, Circuits, Systems, and Signal Processing, Vol. 5, No. 1, 1986.

[2] M.F. Atiyah, I. G. MacDonald, *Introduction to Commutative Algebra*, Addison–Wesley, 1969.

[3] J.W. Brewer, J.W. Bunce, F.S. Van Vleck, *Linear Systems over Commutative Rings*, Marcel Dekker, New York, 1986.

PROBLEMS AND RESULTS IN A GEOMETRIC APPROACH TO THE THEORY OF SYSTEMS OVER RINGS

G. CONTE* AND A.M. PERDON†

To the memory of Mario Raimondo

Abstract. The use of geometric methods in the study of linear systems with coefficients in a ring presents several difficulties, but, as in the case of coefficients in a field, it appears capable of providing efficient solutions to a number of control and observation problems. In this paper we survey some recent results concerning a class of Disturbance Decoupling Problems and we extend the techniques developed in that case to a class of problems concerning the Estimation of Linear Functions of the State.

1. Introduction. The geometric approach to the theory of linear dynamical systems has provided in the last decade deep insights and elegant solutions to many synthesis and control problems, such as the disturbance decoupling problem, the model matching problem, the unknown-input observation problem, the tracking and regulation problem (see [12], [3] and the references therein).

Systems with coefficients in a ring turn out to be useful in many interesting situations for modeling, e.g., delay-differential systems, families of systems depending on one parameter, time-varying systems (see [9], [10], [11]). This motivates the attempts of extending the geometric approach to the theory of systems with coefficients in a ring. The first results in this direction were obtained by M. Hautus and by the authors in 1982 (see [7] and also [8]).

The main difference encountered in dealing with coefficients in a ring instead of coefficients in a field concerns the fundamental notion of (A, B)-invariant subspace or controlled invariant subspace ([1], [12], [3]). From a dynamical point of view, the controlled invariance property for a subspace V of the state space of a system Σ means that the trajectories starting in V can be kept inside V by a suitable choice of the control. What makes the (A, B)-invariant subspaces so useful in the theory of systems with coefficients in a field is the fact that a control with the above property can be generated by using a state feedback. This means, in other terms, that by choosing a suitable state feedback any (A, B)-invariant subspace can be made invariant with respect to the resulting closed loop dynamics. Unfortunately, such crucial feature is generally lost in the framework of systems with coefficients in a ring, where only the so-called (A, B)-invariant submodules of feedback type ([7]) can be made invariant by means of a state

* Dipartimento di Elettronica e Automatica - Università di Ancona - via Brecce Bianche - 60131, Italy.

† Dipartimento di Matematica - Università di Ancona - via Brecce Bianche - 60131, Italy.

feedback. Analogous considerations hold for the notion of (A, C)-invariant subspace or conditioned invariant subspace ([1], [12], [3]). The possibility of making an (A, C)-invariant subspace invariant with respect to a dynamics modified by output injection is generally lost in the framework of systems with coefficient in a ring. Besides, (A, B)-invariance and (A, C)-invariance are not dual when coefficients are taken in a ring.

For the above reasons, the classical results obtained by means of geometric techniques for systems with coefficients in a field can be generalized only partially to systems with coefficients in a ring. A paramount illustration of this is provided in [7] by the Section devoted to the Disturbance Decoupling Problem and in [8]. It turns out that the well-known geometric conditions which characterize the solvability of such problem in the case of systems with coefficients in a field ([1]), [12], [3]) are necessary but not sufficient, except for restricted classes of systems, for assuring the existence of solutions in the case of coefficients in a ring.

A complete characterization in geometric terms of the solvability conditions of the Disturbance Decoupling Problem for a quite general class of systems with coefficients in a ring was only recently given by the authors in [6]. The approach developed in that paper differs slightly from the classical one, since, instead of being based on the properties of some maximal controlled invariant submodule (compare with [12], [3]), it makes use of the minimum elements in particular families of controlled invariant submodules. This approach turns out to be feasible for systems which verify a mild geometric condition, that from a practical point of view is essentially equivalent to injectivity. For systems of this kind, in fact, it is possible to define the smallest controlled invariant submodule containing a given module. In particular, if the coefficients of the system are taken in a Principal Ideal Domain, the results of [6] provide a practical procedure for checking the existence of solutions to a given Disturbance Decoupling Problem and for computing one of them.

The aim of this paper is to survey the approach developed in [6] and to extend it to a class of problems concerning the estimation of linear functions of the state in presence of disturbances. Following the guideline of [6], we discuss, in Section 2, some preliminary aspects of the geometric approach in connection with a class of Disturbance Decoupling Problems. A geometric characterization of the solvability conditions for such problems is described in Section 3, after investigating the properties of particular families of controlled invariant submodules. The relevant computations for checking the solvability conditions and for constructing a solution are described in Section 4 and, for systems with coefficients in a Principal Ideal Domain, in Section 5. In particular, this give us the opportunity of discussing the behavior of the classical geometric algorithms in the framework of systems with coefficients in a ring. In Section 6, which represents the only novelty of this paper with respect to [6], we consider the geometric approach in connection with a class of Estimation Problems. A characterization in geometric terms of the solvability conditions for such problems is given and some computational aspects are described. The last Section contains examples, taken from [6], concerning delay-differential systems and families of parameter dependent systems.

2. (A,B)-invariant submodules and disturbance decoupling problems. Let R denote a commutative ring. By a *system with coefficients in* R, or a *system over* R, we mean a discrete-time linear dynamical system Σ whose evolution is described by a set of difference equations of the form

$$\begin{cases} x(t+1) = Ax(t) + Bu(t) \\ y(t) = Cx(t) \end{cases} \tag{2.1}$$

where x belongs to the free state module $X = R^n$, u belongs to the free input module $U = R^m$, y belongs to the free output module $Y = R^p$, and A, B, C are matrices of suitable dimensions with entries in R.

As a starting point for investigating the possibility of developing a geometric approach for systems over R, let us consider the following control problems

2.1. Statement of the disturbance decoupling problems. Given a system Σ over R, described by the equations

$$\begin{cases} x(t+1) = Ax(t) + Bu(t) + Dq(t) \\ y(t) = Cx(t) \end{cases} \tag{2.2}$$

where $q \in Q = R^q$ is a disturbance, then

i) the Disturbance Decoupling Problem (shortly DDP) consists in finding a state feedback law $u(t) = Fx(t)$, if there exists any, such that the output of the closed loop system Σ_F does not depend on q:

ii) if the disturbance q is assumed to be measurable, the Disturbance Decoupling Problem with Measurable Disturbance (shortly DDPMD) consists in finding a state feedback law $u(t) = Fx(t) + Gq(t)$, if there exists any, such that the output of the compensated system $\Sigma_{F,G}$ does not depend on q.

In the case of linear systems with coefficients in a field, elegant solutions to the above control problems are provided by the so-called geometric approach (see [1], [12], [3]). Early attempts to extend the geometric approach to the framework of the systems over a ring and to apply geometric techniques to the solutions of the DDP and the DDPMD are reported in [7]. In order to describe the results obtained and to discuss the differences and the difficulties encountered in dealing with systems over a ring instead of a field, let us recall the following Definitions.

2.2. Definitions. ([7]) Given a system Σ over R described by (2.1), a submodule V of the state module X is called:

i) (A, B)-invariant, or controlled invariant, if and only if $AV \subset V + ImB$;

ii) (A, B)-invariant of feedback type if and only if there exists an R-linear map $F : X \to U$ such that $(A + BF)V \subset V$.

A feedback F, if any exists, with the property described in ii) above is called a friend of V.

The proof of the following Proposition follows immediately from [1], [12].

2.3. Proposition. Given a system Σ over R described by (2.2), the DDP (respectively the DDPMD) is solvable if and only if there exists an (A, B)-invariant V of feedback type such that $Im\ D \subset V \subset \text{Ker } C$ (respectively such that $Im\ D \subset V + Im\ B$ and $V \subset \text{Ker } C$).

For systems with coefficients in a field, the two notions described in Definition 2.2 are known to coincide. Moreover, the family of the (A, B)-invariant subspaces contained in Ker C has a maximum element, usually denoted by V^*, which can be found as the limit of the sequence of subspaces V_k defined recursively by

$$\text{(2.3)} \qquad \begin{cases} V_0 = \text{Ker } C \\ \\ V_{k+1} = V_k \cap A^{-1}(V_k + Im\ B) \end{cases}$$

Then, the necessary and sufficient condition of Proposition 2.3 can be stated as $Im\ D \subset V^*$ (respectively $Im\ D \subset V^* + Im\ B$) and, since the sequence (2.3) converges in a finite number of steps, it can be easily checked.

For systems with coefficients in a ring, the two notions described in Definition 2.2 do not coincide. In fact, any (A, B)-invariant submodule of feedback type is, in particular, (A, B)-invariant, but the converse is not true. The family of the (A, B)-invariant submodules contained in Ker C has, also in this case, a maximum element V^* (see [7], Proposition 2.5), but the condition $Im\ D \subset V^*$ (respectively $Im\ D \subset V^* + Im\ B$), which is of course necessary for the existence of solutions to the DDP (respectively the DDPMD), is sufficient only if V^* happens to be of feedback type. Unfortunately, this is generally untrue for systems with coefficients in a ring (see [7], in particular Example 5.6). Hence, the geometric conditions which characterize the solvability of the considered problems in the case of coefficients in a field are too weak for providing a complete characterization of the solvability also in the case of coefficients in a ring. Moreover, since there does not exist in general a maximum (A, B)-invariant submodule of feedback type contained in Ker C (see [7], Section 5), the above conditions cannot be modified in a straightforward way.

3. Solvability conditions for the DDP and the DDPMD. A possible strategy for overcoming the difficulties we pointed out in the previous Section consists in characterizing those systems for which V^* turns out to be of feedback type (see [7] and [8]). Systems of this kind form however only a small subset of the class at issue and therefore such attempt does not provide a completely satisfactory result. A different approach, based on a particular (A, B)-invariant submodule, that allows us to state a necessary and sufficient condition for the solvability of the DDP and the DDPMD for injective systems, has been developed in [6]. In order to describe it, let us start by stating some preliminary results about the family of (A, B)-invariant submodules of the state space X of a system Σ described by (2.1).

3.1. Proposition. Given Σ, let $V \subset X$ be an (A, B)-invariant submodule such that $V \cap Im\ B = \{0)\}$. Then, we have:

i) if $V' \subset V$ and $V'' \subset V$ are two (A, B)-invariant submodules, then $V' \cap V''$ is an (A, B)-invariant submodule;

ii) if V is of feedback type, any (A, B)-invariant submodule V' such that $V' \subset V$ is of feedback type; moreover, if $F : X \to U$ is a friend of V, F is also a friend of V';

iii) if $B : U \to X$ is injective, $F : X \to U$ is a friend of V and $F' : X \to U$ is a friend of $V' \subset V$, then $F_{|V'} = F'_{|V'}$.

The above results are a slight generalization of the results about self-bounded invariant subspaces given in [2] for systems with coefficients in a field. The condition $V \cap Im\ B = \{0\}$ implies in fact that the invariant subspaces contained in V are self-bounded with respect to V, i.e., are such that a trajectory starting in one of them, say V', and remaining in V cannot escape from V'. The proofs given in [2] for the field case work also in our situation.

We recall that a system Σ described by (2.1) is said to be injective if for any state x_0 in X and any pair of input sequences $\mathbf{u} = \{u(t),\ 0 \le t\}$, $\mathbf{v} = \{v(t),\ 0 \le t\}$ the equality $\mathbf{y}(x_0, \mathbf{u}) = \mathbf{y}(x_0, \mathbf{v})$ of the corresponding output sequences implies $\mathbf{u} = \mathbf{v}$. If Σ is injective and V^* denotes the maximum (A, B)-invariant submodule contained in Ker C, then $V^* \cap Im\ B = \{0\}$ and B is injective. The results of the above Proposition therefore apply to the (A, B)-invariant submodules contained in Ker C. In particular, by 3.1 i), we have that any family consisting of all the (A, B)-invariant submodules of Ker C that contain a given module has a minimal element. From a practical point of view, injectivity is a very weak assumption, since input sequences that produce the same output can be identified.

3.2. Notation. In the following, given a system Σ for which the condition $V^* \cap Im\ B = \{0\}$ holds and given a family $\mathcal{L}$ of (A, B)-invariant submodules of Ker C, we will denote by $V_*(\mathcal{L})$ the minimal element in the family, provided such element exists.

We can now state the following result, whose proof follows from Proposition 2.3 and Proposition 3.1 ii)

3.3. Theorem. Let Σ be the system described by (2.2) and assume that $V^* \cap Im\ B = \{0\}$, where V^* is the maximum (A, B)-invariant submodule of Ker C. Then, the DDP concerning Σ is solvable if and only if $Im\ D \subset V^*$ and $V_*(\mathcal{L})$, where $\mathcal{L}$ is the family of all the (A, B)-invariant submodules of Ker C containing $Im\ D$, is of feedback type. Analogously, the DDPMD is solvable if and only if $Im\ D \subset V^* + Im\ B$ and $V_*(\mathcal{L})$, where $\mathcal{L}$ is the family of all the (A, B)-invariant submodules V of Ker C such that $Im\ D \subset V + Im\ B$, is of feedback type.

The above Theorem gives a complete geometric characterization of the solvability conditions of the DDP and DDPMD. In order to use it for solving the DDP and the DDPDM, we should be able to construct the submodules V^* and $V_*(\mathcal{L})$, for the family $\mathcal{L}$ we are interested in, and to check whether the latter is of feedback type or not. These problems are considered in the next Section.

4. Properties and computation of $V_*(\mathcal{L})$. The results of Theorem 3.3 assume a practical relevance if a procedure for computing $V_*(\mathcal{L})$ is given. In order to relate $V_*(\mathcal{L})$ to other known geometric objects, let us recall the following Definition.

4.1. Definition. Given a system Σ over R described by (2.1), a submodule S of the state module X is called (A, C)-invariant if and only if $A(S\cap \text{Ker } C) \subset S$.

As in the case of systems with coefficients in a field, the existence of the minimum (A, C)-invariant submodule containing a given submodule $K \subset X$, denoted by $S^*(K)$, is easily proved. If, as in Theorem 3.3, we restrict our attention to systems for which the condition $V^* \cap Im\ B = \{0\}$ holds, we have the following results. (see [2] for the case of systems with coefficients in a field).

4.2. Proposition. Given a system Σ over R described by (2.2) assume that $V^* \cap Im\ B = \{0\}$, where V^* is the maximum (A, B)-invariant submodule of Ker C, and that $Im\ D \subset V^*$. If $\mathcal{L}$ denotes the family of all the (A, B)-invariant submodules of Ker C containing $Im\ D$, then we have $V_*(\mathcal{L}) = V^* \cap S^*(Im\ D + Im\ B)$

Proof. The proof given in the field case in [2] Theorem 2.3 works in our situation.

4.3. Proposition. Given a system Σ over R described by (2.2) assume that $V^* \cap Im\ B = \{0\}$, where V^* is the maximum (A, B)-invariant submodule of Ker C, and that $Im\ D \subset V^* + Im\ B$. If $\mathcal{L}$ denotes the family of all the (A, B)-invariant submodules V of Ker C such that $Im\ D \subset V + Im\ B$, then we have $V_*(\mathcal{L}) = V^* \cap S^*(Im\ D + Im\ B)$.

Proof. See [6], Proposition 4.3.

The computation of $V_*(\mathcal{L})$, for the family $\mathcal{L}$ we are interested in, is therefore reduced to the computation of V^* and of $S^*(Im\ D + Im\ B)$.

Concerning the computations of $S^*(Im\ D + Im\ B)$, let us first remark that, as in the case of coefficients in a field ([1]), for any submodule K of X the sequence of submodules S_k of X defined recursively by

$$(4.1)\qquad \begin{cases} S_0 = K \\ \\ S_{k+1} = S_k + A\ (S_k \cap \text{ Ker } C) \end{cases}$$

is non-decreasing and becomes stationary for $k \geq \overline{k}$ if $S_{\overline{k}+1} = S_{\overline{k}}$. If R is a Noetherian ring we can therefore state that the sequence $\{S_k\}$ converges in a finite number of steps, and moreover we can prove, in the same way as in [1], that the limit of $\{S_k\}$ is $S^*(K)$. The computation of $S^*(K)$ as the limit of $\{S_k\}$ is usually referred to, in the case of coefficients in a field, as the Conditionally Invariant Subspace Algorithm. Then, (4.1) initialized with $S_0 = Im\ D{+}Im\ B$ yields, if R is a Neotherian ring, $S^*(Im\ D{+}Im\ B) = S_{\overline{k}}$, where $\overline{k}$ is the smallest integer for which $S_{\overline{k}+1} = S_{\overline{k}}$.

The situation is quite different for what concerns the computation of V^*. In the case of coefficients in a field, as already mentioned, V^* can be computed as the limit of the sequence of submodules V_k defined recursively by (2.3), that converges in a finite number of steps. This is no longer true for systems over a ring R, since the convergence of $\{V_k\}$ in a finite number of steps is not guaranteed. In the next Section we will give a procedure for finding V^* when R is a Principal Ideal Domain, but, in general, its construction remains a difficult problem.

Finally, for solving a given DDP or DDPDM by means of Theorem 3.3 one needs to check whether the submodule $V_*(\mathcal{L})$ is of feedback type or not. Since no alternative general characterization of property 2.2 ii) is known, this may be a quite difficult task. In the next Section we will show how to accomplish this when the coefficient ring R is a Principal Ideal Domain.

5. DDP and DDPMD for systems over a P.I.D. Let us focus now on the case in which R is a Principal Ideal Domain (P.I.D.). This assumption gives us the possibility of characterizing the (A, B)-invariant submodules of feedback type in a quite simple way, and, moreover, it allows us to find a procedure for the construction of V^*. To this aim, let us first recall the following notions and results, due to the authors, from [4] and [7].

5.1. Definitions. ([4], Definition 1.9) let R be a P.I.D. and let $V \subset X = R^n$ be a submodule. By the closure of V in X we mean the submodule

$$\overline{V} = \{x \in X, \text{ such that } ax \in V \text{ for some } a \neq 0,\ a \in R\}.$$

If V coincides with its closure $\overline{V}$, we say that V is closed in X.

5.2. Remark. It is useful to remark the followings ([4], Proposition 1.10):

i) the closure $\overline{V}$ of V in X is the smallest closed submodule of X containing V;
ii) any submodule V of a finitely generated free module X over a P.I.D. R is closed if and only if it is a direct summand of X;
iii) for any submodule V of X, one has $\dim_R V = \dim_R \overline{V}$.

In addition to Remark 5.2, we point out that the computation of the closure of a submodule V can be performed by computing the Smith factorization of suitable matrix. In fact, if D is a square matrix whose first k columns form a basis of V and whose remaining columns are zero, and if $D = PSQ$, with $S = \text{diag}(a_1, \ldots, a_k, 0, \ldots, 0)$, is a Smith factorization, then the first k columns of $D' = PS'Q$ with $S' = \text{diag}(1, \ldots, 1, 0 \ldots, 0)$ form a basis of $\overline{V}$. Algorithms for constructing a Smith factorization of D are known.

5.3. Proposition. ([7], Proposition 5.3) Let R be a P.I.D. and let V be an (A, B)-invariant submodule of X. Then V is of feedback type only if its closure $\overline{V}$ is (A, B)-invariant.

5.4. Proposition. ([7], Proposition 5.2) Let R be a P.I.D. and let V be a closed (A, B)-invariant submodule of X, then V is of feedback type.

After constructing $V_*(\mathcal{L})$, the notion of closure and Proposition 5.3 gives us the the possibility, when R is a P.I.D., of checking the solvability condition for the DDP and the DDPMD stated in Theorem 3.4. More precisely, Theorem 3.4 becomes

5.5. Theorem. Let Σ be the system described by (2.2) over the P.I.D. R, and assume that $V^* \cap Im\ B = \{0\}$, where V^* is the maximum (A, B)-invariant submodule of Ker C.
Then, the DDP concerning Σ is solvable if and only if $Im\ D \subset V^*$ and the closure $\overline{V_*(\mathcal{L})}$ of $V_*(\mathcal{L})$ in X, where $\mathcal{L}$ is the family of all the (A, B)-invariant submodules of Ker C containing $Im\ D$, is (A, B)-invariant.
Analogously, the DDPMD is solvable if and only if $Im\ D \subset V^* + Im\ B$ and $\overline{V_*(\mathcal{L})}$, where $\mathcal{L}$ is the family of all the (A, B)-invariant submodules V of Ker C such that $Im\ D \subset V + Im\ B$, is (A, B)-invariant.

As a result, we are left now with the problem of constructing V^* in the case in which R is a P.I.D.. To this aim, let us consider the following result.

5.6. Proposition. Given a system Σ described by (2.2) over a P.I.D. R, consider the sequence of submodules $\{V_k\}$ of X defined recursively by (2.3) and denote by $V = \bigcap_{k=0}^{\infty} V_k$ its limit.

i) If $V_k = V_{k+1}$ for some k, then $V_k = V = V^*$, that is the sequence $\{V_k\}$ converges in a finite number of steps to the maximum (A, B)-invariant submodule of Ker C.
ii) If $V_k \neq V_{k+1}$ for all k, then $\overline{V} \subsetneq$ Ker C.

Proof. See [6], Proposition 5.6.

A procedure for constructing V^* in the case in which R is a P.I.D. can now be described as follows:

Step 1) By means of (2.3) compute the sequence $\{V_k\}$ and its limit $V^1 = \bigcap_{k=0}^{\infty} V_k$.
If V^1 is (A, B)-invariant, in particular if $V^1 = V_k$ for some k, $V^1 = V^*$ and the procedure stops.

Step q) In (2.3) set $V_0 = \overline{V^{q-1}}$ and compute the resulting sequence $\{V_k\}$ and its limit $V^q = \bigcap_{k=0}^{\infty} V_k$.
If V^q is (A, B)-invariant, in particular if $V^q = V_k$ for some k, $V^q = V^*$ and the procedure stops.

The key fact to point out in the above procedure is that at each step either the procedure stops or the dimension of the submodule that is used for initializing (2.3) is reduced at least by one. Hence, if $r = \dim$ Ker C, r steps of the above procedure allow us to find V^*. No algorithm for computing the limit of $\{V_k\}$ can obviously be given, but, in general, one may obtain explicit descriptions of V_k that allow to compute the limit.

So, in conclusion, the solvability condition for the DDP and DDPDM for a system over a P.I.D. R can be checked, and a solution can possibly be constructed, by using (4.1) and the above Procedure, 4.2 or 4.3, 5.3 and finally 5.5.

5.7. Remark. Let us assume that for a given system Σ, described by (2.2) over the P.I.D. R, with $V^* \cap Im\ B = \{0\}$, the DDP is solvable with a feedback F. Choosing a basis of X whose first elements are a basis of $\overline{V_*(\mathcal{L})}, \mathcal{L}$ being the family of all the (A, B)-invariant submodules of Ker C containing $Im\ D$, the compensated system S_F is described by the equations

$$(5.1) \qquad \begin{cases} x_1(t+1) = A_{11}x_1(t) + A_{12}x_2(t) + D_1q(t) \\ x_2(t+1) = A_{22}x_2(t) + D_2q(t) \\ y(t) = Cx_2(t) \end{cases}$$

The subsystem $x_1(t+1) = A_{11}x_1(t)$ evolves on $\overline{V_*(\mathcal{L})}$ with a dynamics described, up to a change of basis, by the restriction to such subspace of

$(A+BF)$. The minimality of $V_*(\mathcal{L})$ implies that this is a fixed dynamics for the considered DDP, in the sense that it is present in the compensated system for any solution F. This has important consequences if we modify the DDP or the DDPDM by requiring the additional condition that $\det(zI-A-BF)$ belongs to some specific subset $\mathcal{D}$ of $R[z]$. In particular, $\mathcal{D}$ can be chosen as in [8] for characterizing a notion of internal stability. Then, the DDP or the DDPDM with the additional requirement of internal stability can be solved only if $\det(zI-A_{11})$ belongs to $\mathcal{D}$.

6. (A,C)-invariant submodules and linear function evaluation problems. Let us examine more deeply the properties of the (A,C)-invariant submodules introduced in Definition 4.1. In the case of coefficients in a field, it is well-known ([1]) that, given an (A,C)-invariant subspace S, there exists a linear map $K:Y\to X$ such that $(A+KC)S\subset S$. Let us assume that S is an (A,C)-invariant subspace for the system Σ described by (2.2), having the property that $(A+KC)S\subset S$ for some K and such that $Im\ D\subset S$. Then,

$$z(t+1)=(A+KC)z(t)-Ky(t)+Bu(t) \tag{6.1}$$

is an obvious observer for x. The dynamics of the error $e(t)=x(t)-z(t)$ is governed by

$$e(t+1)=(A+KC)e(t)-Dq(t) \tag{6.2}$$

and therefore, if the initial error $e(0)$ belongs, with a slight abuse of notations, to $S, e(t)$ remains in S. In this case it is possible to compute any linear function of $x(t)$, say $G(x(t))$, such that $S\subset \text{Ker } G$ by $G(x(t))=G(z(t))$, without knowing the disturbance $d(t)$. Unfortunately, as for the (A,B)-invariant submodules, the possibility of making an (A,C)-invariant submodule invariant with respect to a modified dynamics is lost when the coefficients are taken in a ring, even if the ring is a P.I.D. (see Example 7.4). We therefore need the following Definition.

6.1. Definition. Given a system Σ over r described by (2.1), a submodule S of the state module X is called (A,C)-invariant of observer type if and only if there exists an R-linear map $K:Y\to X$ such that $(A+KC)S\subset S$.
A map K, if any exists, with the property described above is called a friend of S.

We can now state more precisely the estimation problem mentioned above and formulate its solvability conditions.

6.2. Statement of the estimation problem for linear functions of the state. Given a system Σ over R described by the equation (2.2), where $q\in Q=R^q$ is a disturbance, and given a linear function $G:X\to$

W, where W is an R-module, then the Estimation Problem for Linear Functions of the State (shortly LFSEP) consists in finding a linear observer for Σ that, using u and y as inputs, provides an estimate $z(t)$ of $x(t)$ verifying $G(x(t)) = G(z(t))$ for $t \geq 0$.

6.3. Proposition. Given a system Σ over R described by (2.2) and given a linear function $G : X \to W$, where W is an R-module, the LFSEP is solvable if and only if there exist an (A, C)-invariant submodule S of observer type such that $Im\ D \subset S \subset$ Ker G and $x(0)$modS is known.

As we have already seen in Section 4, the minimum (A, C)-invariant submodule containing $Im\ D$, namely $S^*(Im\ D)$, can easily be computed and the condition $S^*(Im\ D) \subset$ Ker G can be checked. Unfortunately, $S^*(Im\ D)$ may not be of observer type, and hence the problem at issue presents the same kind of difficulties we have encountered in dealing with the DDP. However, an approach similar to the one used in that case can be easily developed.

6.4. Proposition. Given Σ, let $S \subset X$ be an (A, C)-invariant submodule such that $S+$ Ker $C = X$. Then, we have:

i) if $S' \supset S$ and $S'' \supset S$ are two (A, C)-invariant submodules, then $S' + S''$ is an (A, C)-invariant submodule;

ii) if S is of observer type, any (A, C)-invariant submodule S' such that $S' \supset S$ is of observer type; moreover, if $K : Y \to X$ is a friend of S, K is also a friend of S'.

The above results are a slight generalization of the results about self-hidden invariant subspaces given in [3] for systems with coefficients in a field. A proof of the above results in the field case is usually given by duality with the corresponding situation concerning (A, B)-invariant subspaces. Duality does not work in our case, so we give for completeness a simple direct proof.

Proof of 6.4. i) Let $s_1 + s_2 \in (S' + S'') \cap$ Ker C and write $s_1 = s + k$, with $s \in S$ and $k \in$ Ker C. Since $s + s_2$ turns out to belong to $S'' \cap$ Ker C and k turns out to belong to $S' \cap$ Ker C, we have $A(s_1 + s_2) = A(s + s_2) + A(k) \in S' + S''$.
ii) Let K be a friend of S and for $s_1 \in S'$ write $s_1 = s + k$, with $s \in S$ and $k \in$ Ker C. Since k turns out to belong to $S' \cap$ Ker C, we have $(A + KC)(s_1) = (A + KC)(s) + (A + KC)(k) \in S'$.

Under the hypothesis $S^*(Im\ D)+$ Ker $C = X$, the family $\mathcal{L}$ of all the (A, C)-invariant submodules containing $Im\ D$ and contained in Ker G has a maximum element, that we will denote by $S_*(\mathcal{L})$. Then we can state the following Theorem.

6.5. Theorem. Let Σ be the system described by (2.2) and assume that $S^*(Im\ D)+$ Ker $C = X$. Then, the LFSEP concerning Σ is solvable if and only if $S^*(Im\ D) \subset$ Ker G and $S_*(\mathcal{L})$, where $\mathcal{L}$ is the family of all the (A, C)-invariant submodules of Ker G containing $Im\ D$, is of observer type and $x(0)$ modS is known.

No characterization of the (A, C)-invariant submodules of observer type analogous to that given in Proposition 5.3 for the (A, B)-invariant submodules of feedback type is known, ever when r is a P.I.D.. Hence, the above formulation of the solvability conditions for the LFSEP has only a limited practical applicability and further development, in the spirit of Section 5, are at the moment not possible. It may however be useful to mention the following result.

6.6. Proposition. Let R be a P.I.D.. An (A, C)-invariant submodule S is of observer type if $C(S)$ is closed.

7. Examples.

7.1. Example. Let Σ be the delay-differential system defined by

$$\Sigma = \begin{cases} \dot{x}_1(t) = q(t-2\partial) + q(t-\partial) \\ \dot{x}_2(t) = q(t-2\partial) + 2q(t-\partial) + q(t) \\ \dot{x}_3(t) = x_1(t) + u_1(t-\partial) \\ \dot{x}_4(t) = x_2(t) + u_2(t) \\ \\ y_1(t) = x_3(t) \\ y_2(t) = x_4(t) \end{cases}$$

where ∂ represents a fixed delay and q is a disturbance. By introducing the delay operator Δ, defined for any time function f by $\Delta(f)(t) = f(t-\partial)$, the disturbance decoupling problem for Σ consists in finding a delay feedback law $u(t) = F(\Delta)(x)(t)$, where $F(\Delta)$ is a matrix of polynomials in Δ, such that the output of the compensated system $\Sigma_{F(\Delta)}$ does not depend on q. Formally, we can associate to Σ a system of the form (2.2) defined over the ring of polynomials $R[\Delta]$ by the matrices

$$A = \begin{pmatrix} 0 & 0 & 0 & 0 \\ 0 & 0 & 0 & 0 \\ 1 & 0 & 0 & 0 \\ 0 & 1 & 0 & 0 \end{pmatrix} ; \ B = \begin{pmatrix} 0 & 0 \\ 0 & 0 \\ \Delta & 0 \\ 0 & 1 \end{pmatrix} ; \ D = \begin{pmatrix} \Delta(\Delta+1) \\ (\Delta+1)^2 \\ 0 \\ 0 \end{pmatrix} ;$$

$$C = \begin{pmatrix} 0 & 0 & 1 & 0 \\ 0 & 0 & 0 & 1 \end{pmatrix} .$$

The DDP concerning Σ can then be dealt with using the methods we have developed in the previous Sections. First, an easy computation shows that Ker $C = \{(a, b, 0, 0)^T \ ; \ a, b \in R[\Delta]\}$ and that $V^* = \{(\Delta a, b, 0, 0)^T \ ; \ a, b \in$

$R[\Delta]\}$. Clearly $Im\ D$ is contained in V^*, whose closure $\overline{V^*}$ coincides with Ker C. Since Ker C is not (A,B)-invariant, V^* is not of feedback type and therefore the condition $Im\ D \subset V^*$ is not sufficient for assuring the solvability of the DDP. Remark, moreover, that for this example there does not exists a maximum (A,B)-invariant submodule of feedback type contained in Ker C. In fact, both the submodules $V_1 = \{(\Delta\ a, (\Delta + 1)\ a,\ 0,\ 0)^T;\ a \in R[\Delta]\}$ and $V_2 = \{(\Delta^2 a, (\Delta+1)a, 0, 0)^T\ ;\ a \in R[\Delta]\}$ are (A,B)-invariant and closed. Hence, the maximum (A,B)-invariant submodule of feedback type contained in Ker C, if it exists, must contain both V_1 and V_2. Then, since it may be assumed to be closed, it must contain also $\overline{V_1 + V_2}$. But, as $\overline{V_1 + V_2}$ coincides with Ker C, no maximum exists. A simple computation shows that $Im\ D$ is (A,B)-invariant, hence $V_*(Im\ D) = Im\ D$. Since $D_1 = \overline{V_*(Im\ D)} = \{(\Delta a, (\Delta+1)a, 0, 0)^T\ ;\ a \in R[\Delta]\}$ is (A,B)-invariant, $V_*(Im\ D)$ is of feedback type and there exists a solution to the considered DDP. In order to compute such solution, let us extend the basis $\{(\Delta, \Delta+1, 0, 0)^T\}$ of $\overline{V_*(Im\ D)}$ to a basis $\{(\Delta, \Delta+1, 0, 0)^T, (1,1,0,0)^T, (0,0,1,0)^T, (0,0,0,1)^T\}$ of $(R[\Delta]^4$ and, remarking that $A\ ((\Delta, \Delta+1, 0, 0)^T) = B((1, \Delta+1)^T)$, let define a map $f : (R[\Delta])^4 \to (R[\Delta])^2$ by $f((\Delta, \Delta+1, 0, 0)^T) = (1, \Delta+1)^T$, and e.g. $f((1,1,0,0)^T) = f((0,0,1,0)^T) = f((0,0,0,1)^T) = (0,0)^T$. The matrix F associated to f with respect to the canonical basis is

$$F = \begin{pmatrix} -1 & 1 & 0 & 0 \\ -\Delta - 1 & \Delta + 1 & 0 & 0 \end{pmatrix}.$$

The compensated system Σ_F takes now the following form

$$\Sigma = \begin{cases} \dot{x}_1(t) = q(t-2\partial) + q(t-\partial) \\ \dot{x}_2(t) = q(t-2\partial) + 2q(t-\partial) + q(t) \\ \dot{x}_3(t) = x_1(t) + x_1(t-\partial) - x_2(t-\partial) \\ \dot{x}_4(t) = x_1(t) + x_1(t-\partial) - x_2(t-\partial) \\ \\ y_1(t) = x_3(t) \\ y_2(t) = x_4(t) \end{cases}$$

from which one can check that the second derivative $\ddot{y}(t)$ of the output $y(t)$ is equal to zero, thus proving that $y(t)$ is actually independent of q.

7.2. Example. Let $\Sigma(\mathbf{p})$ be the family of parameter dependent systems defined by

$$\Sigma = \begin{cases} \dot{x}_1(t) = \mathbf{p}(\mathbf{p}+1)q(t) \\ \dot{x}_2(t) = (\mathbf{p}+1)^2 q(t) \\ \dot{x}_3(t) = x_1(t) + \mathbf{p}u_1(t) \\ \dot{x}_4(t) = x_2(t) + u(t) \\ \\ y_1(t) = x_3(t) \\ y_2(t) = x_4(t) \end{cases}$$

where $\mathbf{p}$ represents a parameter which may take value in a fixed set $\mathbf{P}$ and q is a disturbance. The disturbance decoupling problem for Σ we are interested in solving consists in finding a feedback law $u(t) = F(\mathbf{p})(x)(t)$, where $F(\mathbf{p})$ is a matrix that depends polynomially on $\mathbf{p}$, such that the output of the compensated system $\Sigma_{F(\mathbf{p})}$ does not depend on q. Formally, we can associate to Σ a system of the form (2.2) defined over the ring of polynomials $R[\mathbf{p}]$ by the same matrices A, B, D, C as in Example 7.1, with Δ replaced by $\mathbf{p}$. The conclusion follows as in Example 7.1 and the following solution is found

$$F = \begin{pmatrix} -1 & 1 & 0 & 0 \\ -\mathbf{p}-1 & \mathbf{p}+1 & 0 & 0 \end{pmatrix} .$$

7.3. Example. Let $R = R[X]$ and let $A = \begin{pmatrix} 0 & 1 \\ 1 & 0 \end{pmatrix}$, $C = (X \quad 0)$. The submodule S of R^2 generated by $e_1 = \begin{pmatrix} 1 \\ 0 \end{pmatrix}$ is a (A, C)-invariant, but no $K = \begin{pmatrix} \alpha \\ \beta \end{pmatrix}$, with α, β belonging to r, exists such that $A(+KC)S \subset S$, since this would imply $1 + \beta X = 0$.

REFERENCES

[1] G. Basile and G. Marro, *Controlled and Conditioned Invariant Subspaces in Linear System Theory*, J.O.T.A. 3 (1969).

[2] G. Basile and G. Marro, *Self-bounded controlled invariant subspaces: a straightforward approach to constrained controllability*, J.O.T.A. 38 (1982).

[3] G. Basile and G. Marro, *Controlled and Conditioned Invariants in Linear System Theory*, Prentice Hall, 1991.

[4] G. Conte and A.M. Perdon, *Systems over Principal Ideal Domains. A Polinomial Model Approach*, SIAM J. Control Opt. 20 (1982).

[5] G. Conte and A.M. Perdon, *The disturbance decoupling problem for systems over a principal ideal domain*, New Trends in systems Theory, G. Conte, A.M. Perdon and B. Wyman Eds., Birkhauser (1991).

[6] G. Conte and A.M. Perdon, *The Disturbance Decoupling Problem for Systems over a Ring*, in preparation.

[7] M.L.J. Hautus, *Controlled invariance in systems over rings*, Springer Lecture Notes in Control and Info. Sci. 39 (1982).

[8] J.L.J. Hautus, *Disturbance rejection for systems over rings*, Springer Lecture Notes in Control and Info. Sci. 58 (1984).

[9] E. Kamen, *Lectures on Algebraic System Theory. Linear Systems over Rings*, NASA Contractor Report 3016 (1976).

[10] E. Sontag, *Linear Systems over Rings: a Survey*, Ricerche di Automatica 7 (1976).

[11] E. Sontag, *Linear Systems over Rings: a (Partial) Update Survey*, Proc. IFAC/81, Kyoto,(1981).

[12] M. Wohnam, *Linear multivariable control: a geometric approach*, 3rd Ed. Springer Verlag, 1985.

COMPLETION OF A MATRIX SO THAT THE INVERSE HAS MINIMUM NORM. APPLICATION TO THE REGULARIZATION OF DESCRIPTOR CONTROL PROBLEMS

L. ELSNER*, C. HE†, AND V. MEHRMANN†

Abstract. We discuss the problem of minimizing the spectral norm of the inverse of a matrix, when a submatrix of the matrix can be chosen arbitrarily. In a recent paper by the authors [3], it was shown that the solution of this problem can be discussed in a similar way as the problem of minimizing the norm of the matrix in terms of matrix Riccati inequalities.

Here we review the results for the norm of the inverse and then apply these results to the robust regularization of descriptor control problems. We also describe a numerical method and give numerical examples.

1. Introduction. We study the following optimization problem: Given integers $n, m, N > n, m$ and matrices $A \in \mathbb{C}^{n,m}, B \in \mathbb{C}^{n,N-m}, C \in \mathbb{C}^{N-n,m}$, find $X \in \mathbb{C}^{N-n,N-m}$ such that the matrix

$$W(X) = \begin{bmatrix} A & B \\ C & X \end{bmatrix} \tag{1.1}$$

satisfies

$$\|W(X)^{-1})\| = \min_{Z \in \mathbb{C}^{N-n,N-m}} \{\|W(Z)^{-1}\|\}. \tag{1.2}$$

Throughout the paper $\|\ \|$ will denote the spectral norm and N the order of $W(X)$. Here we include the case that $m = 0$. (The case $n = 0$ would follow in a similar way by transposition.)

The analysis of this problem is motivated by the following problem of robust regularization of descriptor control problems, which was introduced in Bunse–Gerstner/Mehrmann/Nichols [1,2].

Consider the linear descriptor control problem:

$$E\dot{x} = Fx + Gu, \ y = Hx \tag{1.3}$$

with $E, F \in \mathbb{C}^{n_1,n_1}, G \in \mathbb{C}^{n_1,m_1}, H \in \mathbb{C}^{p_1,n_1}$. Here x is the state, u the input and y the output of the system and E is assumed to be singular. Without loss of generality let us assume that

$$E = \begin{bmatrix} \Sigma & 0 \\ 0 & 0 \end{bmatrix}, F = \begin{bmatrix} F_{11} & F_{12} \\ F_{21} & F_{22} \end{bmatrix}, G = \begin{bmatrix} G_1 \\ G_2 \end{bmatrix}, H = \begin{bmatrix} H_1 & H_2 \end{bmatrix}, \tag{1.4}$$

* Fakultät für Mathematik, Universität Bielefeld, Postfach 8640, D–4800 Bielefeld, Germany.

† Fachbereich Mathematik, TU Chemnitz-Zwickau, Reichenhainer Str.41, D-09009 Chemnitz, Germany.

with Σ square, nonsingular and diagonal. This can easily be achieved via a singular value decomposition of E. If F_{22} is nonsingular, then problem (1.3) can be reduced to a lower order ordinary control system, by eliminating the second block row. In such a case the system is called *regular, index 1.* For details see [1,2].

With x partitioned as $\begin{bmatrix} x_1 \\ x_2 \end{bmatrix}$, we obtain the reduced order system

$$
\begin{aligned}
\Sigma\dot{x}_1 &= (F_{11} - F_{12}F_{22}^{-1}F_{21})x_1 + (G_1 - F_{12}F_{22}^{-1}G_2)u, \\
y &= (H_1 - H_2F_{22}^{-1}F_{12})x_1 - H_2F_{22}^{-1}G_2u
\end{aligned} \tag{1.5}
$$

Using a linear output feedback $u = My$ in (1.3), we can modify the properties of the matrix F. In particular under some regularizability assumptions, (see [2,1]), we can choose the feedback matrix M to make $F_{22} + G_2MH_2$ nonsingular. If we have done so, we can in principle transform the system to the reduced order ordinary control system (1.5). To do this in practice, however, we need that the matrix F_{22} is well-conditioned with respect to inversion. Thus, it is obvious that we should choose the feedback M such that $F_{22} + G_2MH_2$ is well-conditioned. How to do this in general, however, is an open problem. The only case where a complete answer is known is if either H_2 or G_2 is the identity matrix. This arises for example if we use state feedback instead of output feedback. In this case a partial solution was obtained in [2] and a complete characterization is given in [3]. In the general case we can minimize upper bounds for this condition number in different ways, see [1,2,3]). Certainly a good choice for an upper bound would be the minimization of the norm of the inverse of $F_{22} + G_2MH_2$ and for this problem we give a complete characterization of the solution as well as a numerical method. We discuss this problem separately in the Hermitian case, i.e. $A = A^H$, $C = B^H$ and $X = X^H$, in the non-Hermitian case and in the case that $m = 0$.

We use the following notation: By $\mathbb{C}^{n,m}$ we denote the set of complex $n \times m$ matrices. For $A \in \mathbb{C}^{n,m}$ we denote by A^H the conjugate transpose of A. For Hermitian matrices $A, B \in \mathbb{C}^{n,n}$ we write $A < B(A \leq B)$ if $B - A$ is positive definite (positive semidefinite). For a matrix $A \in \mathbb{C}^{n,m}$, we denote by $0 \leq \sigma_p(A) \leq \sigma_{p-1}(A) \leq \ldots \leq \sigma_1(A)$ the singular values of A, where $p = min\{m, n\}$.

2. Minimizing the norm of the inverse. We first discuss the Hermitian case. Let

$$
W(X) = \begin{bmatrix} A & B \\ B^H & X \end{bmatrix}, \tag{2.1}
$$

where A is Hermitian and X is required to be Hermitian, too. The main idea for the solution of problem (1.2) is to consider the limiting case for the solution of the following problem:

For $0 < \alpha < \sigma_n([A, B])$, find an X such that

$$\|W(X)^{-1}\| \leq \frac{1}{\alpha}. \tag{2.2}$$

Here we assume that $0 < \sigma_n([A, B])$ in order to guarantee that $W(X)$ is invertible for some X. In [3] the following results were proved:

THEOREM 2.1. *[3] Let* $W(X) = \begin{bmatrix} A & B \\ B^H & X \end{bmatrix}$ *with A Hermitian.*

(i) Let $0 < \sigma_n([A, B])$. If $0 < \alpha < \sigma_n([A, B])$ and α is not an eigenvalue of A, then

$$\|W(X)^{-1}\| \leq \frac{1}{\alpha} \tag{2.3}$$

always has a solution.

(ii) The set of solutions of (2.3) is equal to the set of solutions of the Riccati inequality $R(X, \alpha) \geq 0$, where

$$R(X,\alpha) = (B^H B + X^2 - \alpha^2 I) - (B^H A + X B^H)(A^2 + BB^H - \alpha^2 I)^{-1}(AB + BX) \tag{2.4}$$

(iii) There are two particular solutions

$$\begin{aligned} Y_1 &= \alpha I + B^H(A - \alpha I)^{-1}B, \\ Y_2 &= -\alpha I + B^H(A + \alpha I)^{-1}B, \end{aligned} \tag{2.5}$$

and general solutions are of the form

$$X = Y_1 + Y \tag{2.6}$$

or

$$X = Y_2 - Y, \tag{2.7}$$

where Y satisfies

$$Y \geq Y(\frac{B^H(A^2 + BB^H - \alpha^2 I)^{-1}B - I}{2\alpha})Y. \tag{2.8}$$

(iv)

$$\inf_X \|W(X)^{-1}\| = \frac{1}{\sigma_n([A, B])}.$$

(v) If $\sigma_n([A, B])$ is not an eigenvalue of A or $-A$, then there exists an Hermitian matrix X such that

$$\|W(X)^{-1}\| = \frac{1}{\sigma_n([A, B])}.$$

REMARK 2.1.

For $R(X,\alpha) \geq 0$, the two special solutions Y_1 and Y_2 do not have a natural order, as

$$Y_1 - Y_2 = 2\alpha R_\alpha, \tag{2.9}$$

where

$$R_\alpha := I - B^H(AA^H + BB^H - \alpha^2 I)^{-1}B \tag{2.10}$$

is indefinite. Only in the case that $0 < \alpha < \sigma_n(A)$, we have that R_α is positive definite and

$$Y_2 \leq Y_1$$

holds, see [3].

If Y is chosen to be nonsingular, then (2.8) implies that

$$Y^{-1} \geq \frac{B^H(A^2 + BB^H - \alpha^2 I)^{-1}B - I}{2\alpha} = -\frac{R_\alpha}{2\alpha}. \tag{2.11}$$

In particular when $\alpha < \sigma(A)$, then R_α is positive definite and $Y_2 \leq Y_1$, and thus any positive definite matrix Y will satisfy (2.8). Thus, in this case we have that the solutions of (2.2) lie outside of $Y_2 \leq Y \leq Y_1$.

REMARK 2.2. When we take the limit $\alpha \to \sigma_n([A,B])$ with $\alpha < \sigma_n([A,B])$,

$$\|W(X)^{-1}\| = \frac{1}{\sigma_n([A,B])}$$

may not have a finite solution. An example is given in [3]. So solving (1.2) requires as an extra condition, that $\sigma_n([A,B])$ is not an eigenvalue of A or $-A$.

We now discuss the problem of minimizing the norm of the inverse of

$$W(X) = \begin{bmatrix} A & B \\ C & X \end{bmatrix}, \tag{2.12}$$

where $A \in \mathbb{C}^{n,m}$, $B \in \mathbb{C}^{n,N-m}$, $C \in \mathbb{C}^{N-n,m}$, and $X \in \mathbb{C}^{N-n,N-m}$.

Let in the following

$$\alpha_0 = \min\{\sigma_n([A,B]), \sigma_m(\begin{bmatrix} A \\ C \end{bmatrix})\}. \tag{2.13}$$

We furthermore assume that $0 < \alpha_0$ in order to guarantee that the inverse of $W(X)$ exists.

The results of the Hermitian case are easyly extended to the more general problem (1.2). As before we solve the problem

$$\|W(X)^{-1}\| \leq \frac{1}{\alpha}, \tag{2.14}$$

for $0 < \alpha < \alpha_0$, and then take limits for $\alpha \to \alpha_0$.

Define the matrices

$$
\begin{aligned}
R_{\alpha,B} &= I - B^H(AA^H + BB^H - \alpha^2 I)^{-1}B \\
R_{\alpha,C} &= I - C(A^H A + C^H C - \alpha^2 I)^{-1}C^H
\end{aligned}
\tag{2.15}
$$

with inverses

$$
\begin{aligned}
R_{\alpha,B}^{-1} &= I + B^H(AA^H - \alpha^2 I)^{-1}B, \\
R_{\alpha,C}^{-1} &= I + C(A^H A - \alpha^2 I)^{-1}C^H.
\end{aligned}
\tag{2.16}
$$

THEOREM 2.2. *[3] Suppose that* $0 < \alpha_0$.
(i) Let $0 < \alpha < \alpha_0$ *and assume that* α *is not an eigenvalue of* A. *Then, the inequality*

$$\|W(X)^{-1}\| \le \frac{1}{\alpha} \tag{2.17}$$

has a solution.
(ii) If $0 < \alpha < \alpha_0$, *then problem (2.14) and the Riccati inequality*

$$
\begin{aligned}
R(X,\alpha) = (CC^H + XX^H - \alpha^2 I) - \\
(CA^H + XB^H)(AA^H + BB^H - \alpha^2 I)^{-1}(AC^H + BX^H) \ge 0.
\end{aligned}
\tag{2.18}
$$

have the same set of solutions.
(iii) All matrices X *solving inequality (2.17) are given by*

$$X = CA^H(AA^H - \alpha^2 I)^{-1}B + \alpha Y, \tag{2.19}$$

where Y *satisfies the inequality*

$$YR_{\alpha,B}Y^H \ge R_{\alpha,C}^{-1}. \tag{2.20}$$

(iv)

$$\inf_X \|W(X)^{-1}\| = \frac{1}{\alpha_0}. \tag{2.21}$$

(v) Furthermore, if we assume that α_0 *is not a singular value of* A, *then there exists a matrix* X *such that*

$$\|W(X)^{-1}\| = \frac{1}{\alpha_0}.$$

Analogous remarks as the ones following Theorem 2.1 hold here, too.

As a corollary we also have the solution for the case that $m = 0$, i.e.

$$W(X) = \begin{bmatrix} B \\ X \end{bmatrix}.$$

COROLLARY 2.3. *Suppose that* $0 < \alpha_0 = \sigma_n(B)$.
(i) Let $0 < \alpha < \alpha_0$. *Then, the inequality*

$$\|W(X)^{-1}\| \leq \frac{1}{\alpha} \tag{2.22}$$

has a solution.
(ii) If $0 < \alpha < \alpha_0$, *then problem (2.14) and the Riccati inequality*

$$R(X,\alpha) = X(I - B^H(BB^H - \alpha^2 I)^{-1}B)X^H - \alpha^2 I \geq 0. \tag{2.23}$$

have the same set of solutions.
(iii) All matrices X *solving inequality (2.23) are given by*

$$X = \alpha Y, \tag{2.24}$$

where Y *satisfies the inequality*

$$Y R_{\alpha,B} Y^H \geq I, \tag{2.25}$$

where in this case $R_{\alpha,B} = I - B^H(BB^H - \alpha^2 I)^{-1}B$.
(iv)

$$\inf_X \|W(X)^{-1}\| = \frac{1}{\alpha_0}.$$

(v) There exists a matrix X *such that*

$$\|W(X)^{-1}\| = \frac{1}{\alpha_0}.$$

(vi) A particular solution is $X = \alpha_0 Y = \alpha_0 \begin{bmatrix} 0 & I_{N-n} \end{bmatrix} V^H$, *where* V *is the right factor of the singular value decomposition* $B = U\Sigma V^H$.

Proof. Parts (i) through (v) follow directly from Theorem 2.2, while part (vi) follows directly from (2.20) by taking the singular value decomposition of B. □

Observe that in the case discussed in Corollary 2.3 the more difficult problem of minimizing the condition number is also solved in [3].

An important result that is needed for the solution of the matrix inequality (2.20) is the following result on the inertia of the matrices $R_{\alpha,B}$ and $R_{\alpha,C}$.

THEOREM 2.4. *Let* $0 < \alpha_0$ *and assume that* α_0 *is not a singular value of* A. *If the inertia of* $AA^H - \alpha_0^2 I$ *is* $(p, n-p, 0)$ *then the number of positive eigenvalues of both* $R_{\alpha_0,B}$ *and* $R_{\alpha_0,C}$ *is equal to* $N + p - n - m$.

Proof. The proof follows in a similar way as the proof of Lemma 2 in [3]. Using Schur complements, e.g. [5], we obtain the congruences

$$\begin{aligned} &\begin{bmatrix} -(AA^H + BB^H - \alpha_0^2 I) & 0 \\ 0 & I_{N-m} \end{bmatrix} \\ &\approx \begin{bmatrix} -(AA^H - \alpha_0^2 I) & B \\ B^H & I_{N-m} \end{bmatrix} \approx \begin{bmatrix} -(AA^H - \alpha_0^2 I) & 0 \\ 0 & R_{\alpha_0,B}^{-1} \end{bmatrix} \end{aligned} \tag{2.26}$$

$$\begin{aligned}(2.27)\quad &\begin{bmatrix} -(A^HA + C^HC - \alpha_0^2 I) & 0 \\ 0 & I_{N-n} \end{bmatrix} \\ &\approx \begin{bmatrix} -(A^HA - \alpha_0^2 I) & C^H \\ C & I_{N-n} \end{bmatrix} \approx \begin{bmatrix} -(A^HA - \alpha_0^2 I) & 0 \\ 0 & R_{\alpha_0,C}^{-1} \end{bmatrix}\end{aligned}$$

we obtain immediately that both $R_{\alpha_0,B}^{-1}$ and $R_{\alpha_0,C}^{-1}$ have $N+p-n-m$ positive eigenvalues. □

3. Application to the control problem. In this section we apply the results of the previous section to the descriptor control problem

$$(3.1)\qquad E\dot{x} = Fx + Gu, \ y = Hx$$

with $E, F \in \mathbb{C}^{n_1,n_1}, G \in \mathbb{C}^{n_1,m_1}, H \in \mathbb{C}^{p_1,n_1}$. Using the first 4 steps of a reduction procedure described in [1], which is essentially a sequence of singular value decompositions, we obtain that (3.1) is equivalent to

$$(3.2)\qquad \tilde{E}\dot{z} = \tilde{F}z + \tilde{G}\tilde{u}, \ \tilde{y} = \tilde{H}x,$$

where

$$(3.3)\qquad \begin{aligned}\tilde{E} &= \begin{bmatrix} \Sigma_E & 0 & 0 & 0 \\ 0 & 0 & 0 & 0 \\ 0 & 0 & 0 & 0 \\ 0 & 0 & 0 & 0 \end{bmatrix}, \tilde{F} = \begin{bmatrix} F_{11} & F_{12} & F_{13} & F_{14} \\ F_{21} & F_{22} & F_{23} & F_{24} \\ F_{31} & F_{32} & \Sigma_F & 0 \\ F_{41} & F_{42} & 0 & 0 \end{bmatrix}, \\ \tilde{G} &= \begin{bmatrix} G_{11} & G_{12} \\ \Sigma_G & 0 \\ 0 & 0 \\ 0 & 0 \end{bmatrix}, \tilde{H} = \begin{bmatrix} H_{11} & \Sigma_H & 0 & 0 \\ H_{21} & 0 & 0 & 0 \end{bmatrix}\end{aligned}$$

are partitioned accordingly and $\Sigma_E, \Sigma_F, \Sigma_G, \Sigma_H$ are square, nonsingular diagonal matrices of sizes t_1, t_3, t_2, s_2, respectively. Set $s_4 := n_1 - t_1 - s_2 - t_3$, $t_4 := n_1 - t_1 - t_2 - t_3$.

We want to find a feedback $\tilde{u} = M\tilde{y} + v = \begin{bmatrix} M_{11} & M_{12} \\ M_{21} & M_{22} \end{bmatrix} \tilde{y} + v$, M partitioned in accordance with $\tilde{G}$ and $\tilde{H}$, such that the closed loop system

$$(3.4)\qquad \tilde{E}\dot{z} = (\tilde{F} + \tilde{G}M\tilde{H})z + \tilde{G}v, \ \tilde{y} = \tilde{H}z$$

is regular, index 1 and the norm of the inverse of the lower right $(n_1 - t_1) \times (n_1 - t_1)$ block of $\tilde{F} + \tilde{G}M\tilde{H}$ is minimal.

To do this, we perform singular value decompositions

$$U^H \begin{bmatrix} F_{32} & \Sigma_F & 0 \\ F_{42} & 0 & 0 \end{bmatrix} V = \begin{bmatrix} \Gamma & 0 \end{bmatrix} = \begin{bmatrix} \gamma_1 & & & 0 & \dots & 0 \\ & \ddots & & \vdots & & \vdots \\ & & \gamma_{t_3+t_4} & 0 & \dots & 0 \end{bmatrix},$$

$$\tilde{U}^H \begin{bmatrix} F_{23} & F_{24} \\ \Sigma_F & 0 \\ 0 & 0 \end{bmatrix} \tilde{V} = \begin{bmatrix} \Delta \\ 0 \end{bmatrix} = \begin{bmatrix} \delta_1 & & \\ & \ddots & \\ & & \delta_{t_3+s_4} \\ 0 & \dots & 0 \\ \vdots & & \vdots \\ 0 & \dots & 0 \end{bmatrix}$$

(3.5)

and obtain $\alpha_0 = \min\{\gamma_{t_3+t_4}, \delta_{t_3+s_4}\}$.

If either $\alpha_0 = 0$ or $s_4 > t_2$ or $t_4 > s_2$ then it is not possible to find a feedback that makes the problem regular, index 1. We assume in the following that this is not the case, so we have Γ, Δ nonsingular.

We furthermore assume that α_0 is not a singular value of Σ_F. If this is the case, then we choose $\alpha < \alpha_0$, such that α is not too close to a singular value of Σ_F, and determine X as below with α_0 replaced by this α.

We can then apply the results from the previous section with

$$A = \begin{bmatrix} \Sigma_F & 0 \\ 0 & 0 \end{bmatrix}, \; B = \begin{bmatrix} F_{32} \\ F_{42} \end{bmatrix} C = \begin{bmatrix} F_{23} & F_{24} \end{bmatrix}, \; X = F_{22} + \Sigma_G M_{11} \Sigma_H.$$

(3.6)

By Theorem 2.2 we obtain that

$$X = \begin{bmatrix} F_{23} & F_{24} \end{bmatrix} \begin{bmatrix} \Sigma_F & 0 \\ 0 & 0 \end{bmatrix} \left(\begin{bmatrix} \Sigma_F^2 & 0 \\ 0 & 0 \end{bmatrix} - \alpha_0^2 I \right)^{-1} \begin{bmatrix} F_{32} \\ F_{42} \end{bmatrix} + \alpha_0 Y,$$

(3.7)

where Y satisfies the inequality

$$Y R_{\alpha_0,B} Y^H \geq R_{\alpha_0,C}^{-1} \tag{3.8}$$

with

$$R_{\alpha_0,B}^{-1} = I + \begin{bmatrix} F_{32} \\ F_{42} \end{bmatrix}^H \left(\begin{bmatrix} \Sigma_F^2 & 0 \\ 0 & 0 \end{bmatrix} - \alpha_0^2 I \right)^{-1} \begin{bmatrix} F_{32} \\ F_{42} \end{bmatrix} \tag{3.9}$$

and

$$R_{\alpha,C}^{-1} = I + \begin{bmatrix} F_{23} & F_{24} \end{bmatrix} \left(\begin{bmatrix} \Sigma_F^2 & 0 \\ 0 & 0 \end{bmatrix} - \alpha_0^2 I \right) \begin{bmatrix} F_{23}^H \\ F_{24}^H \end{bmatrix} \tag{3.10}$$

For any Y satisfying (3.8) we then have with

$$M_{11} = \Sigma_G^{-1} (X - F_{22}) \Sigma_H^{-1}, \tag{3.11}$$

and M_{12}, M_{21}, M_{22} arbitrary (so for simplicity we choose these to be 0), that

$$\tilde{F} + \tilde{G}M\tilde{H} = \begin{bmatrix} F_{11} + G_{11}M_{11}H_{11} & F_{12} + G_{11}M_{11}\Sigma_H & F_{13} & F_{14} \\ F_{21} + \Sigma_G M_{11}H_{11} & F_{22} + \Sigma_G M_{11}\Sigma_H & F_{23} & F_{24} \\ F_{31} & F_{32} & \Sigma_F & 0 \\ F_{41} & F_{42} & 0 & 0 \end{bmatrix}, \tag{3.12}$$

and the norm of the inverse of

$$\begin{bmatrix} F_{22} + \Sigma_G M_{11}\Sigma_H & F_{23} & F_{24} \\ F_{32} & \Sigma_F & 0 \\ F_{42} & 0 & 0 \end{bmatrix}$$

is minimal.

It remains to solve (3.8). Now observe that by our assumption that α_0 is not a singular value of A, we have that both matrices $R^{-1}_{\alpha_0,B}$ and $R^{-1}_{\alpha_0,C}$ exist. Thus, we can perform spectral factorizations or LDL^H factorizations of these matrices, e.g. [4],

$$R^{-1}_{\alpha_0,B} = L \begin{bmatrix} D_1 & 0 \\ 0 & D_2 \end{bmatrix} L^H \tag{3.13}$$

$$R^{-1}_{\alpha_0,C} = \tilde{L} \begin{bmatrix} \tilde{D}_1 & 0 \\ 0 & \tilde{D}_2 \end{bmatrix} \tilde{L}^H \tag{3.14}$$

with $L, \tilde{L}$ nonsingular, $D_1, \tilde{D}_1$ diagonal and positive definite and $D_2, \tilde{D}_2$ diagonal and negative semidefinite. By Theorem 2.4 we have that D_1 and $\tilde{D}_1$ have the same size. Set

$$\tilde{Y} = \tilde{L}^{-1} Y L^{-H} = \begin{bmatrix} Y_{11} & Y_{12} \\ Y_{21} & Y_{22} \end{bmatrix}, \tag{3.15}$$

then (3.8) is equivalent to

$$\tilde{Y} \begin{bmatrix} D_1^{-1} & 0 \\ 0 & D_2^{-1} \end{bmatrix} \tilde{Y}^H \geq \begin{bmatrix} \tilde{D}_1 & 0 \\ 0 & \tilde{D}_2 \end{bmatrix}. \tag{3.16}$$

If we choose $Y_{11} = \hat{D}_1 = (D_1\tilde{D}_1)^{1/2}$ and Y_{12}, Y_{21}, Y_{22} to be zero matrices of appropriate sizes, then (3.16) is satisfied. Note that here we allow that both D_2 and $\tilde{D}_2$ may be singular.

The computational complexity of this procedure can be reduced significantly if we make use of the already computed singular value decompositions (3.5), but since this procedure is rather technical to describe, we refrain here from presenting it.

4. Numerical results. We have implemented the described procedure for the regularization of descriptor control problem in a 368-MATLAB code that is available from the authors. In this section we present several representative numerical examples.

EXAMPLE 4.1. Let

$$E = \begin{bmatrix} 1 & 0 & 0 & 0 \\ 0 & 0 & 0 & 0 \\ 0 & 0 & 0 & 0 \\ 0 & 0 & 0 & 0 \end{bmatrix}, \quad F = \begin{bmatrix} 1000 & 0 & 0 & 0 \\ 0 & 0 & 0 & 0 \\ 0 & 0 & 1000 & 0 \\ 0 & 0 & 0 & 10 \end{bmatrix},$$

$$G = \begin{bmatrix} 0 & 0 \\ 1 & 0 \\ 0 & 1 \\ 0 & 0 \end{bmatrix}, \quad H = G^T.$$

This system is already in the reduced form (3.3) and $\lambda E - F$ is a singular pencil. We obtain $\alpha_0 = 10$, $X = 9.9999 I_2$ and

$$\begin{bmatrix} F_{22} + \Sigma_G M_{11} \Sigma_H & F_{23} & F_{24} \\ F_{32} & \Sigma_F & 0 \\ F_{42} & 0 & 0 \end{bmatrix} = 1000 I_3,$$

which has condition number 1 and norm of the inverse 0.1.

EXAMPLE 4.2. Let E, G, H be as in Example 4.1 and

$$F = \begin{bmatrix} 1000 & 0 & 0 & 0 \\ 0 & 0 & 0 & 0 \\ 0 & 0 & 1000 & 0 \\ 0 & 0 & 0 & 1000 \end{bmatrix}.$$

This system is also in the reduced form (3.3) and $\lambda E - F$ is a singular pencil. We obtain that $\alpha_0 = 1000$ is a singular value of Σ_F. Choosing $\alpha = \alpha_0 + .0001$ we obtain $X = 999.9999 I_2$,

$$\begin{bmatrix} F_{22} + \Sigma_G M_{11} \Sigma_H & F_{23} & F_{24} \\ F_{32} & \Sigma_F & 0 \\ F_{42} & 0 & 0 \end{bmatrix} = 10 I_3,$$

which has condition number 1 and norm of the inverse 0.001.

EXAMPLE 4.3. Let E, F, H be as in Example 4.1 and

$$G = \begin{bmatrix} 0 & 0 \\ 1 & 0 \\ 0 & 0 \\ 0 & 0 \end{bmatrix}$$

This system is also in the reduced form (3.3) and $\lambda E - F$ is a singular pencil. Here $\alpha_0 = 0$ and hence it is not possible to make the pencil regular, index 1 by output feedback.

EXAMPLE 4.4. Let E, G be as in Example 4.3 and

$$F = \begin{bmatrix} 1000 & 0 & 0 & 0 \\ 0 & 10 & -1 & 1 \\ 0 & 0 & 1000 & -2 \\ 0 & 7 & 0 & 1 \end{bmatrix}, \quad H = \begin{bmatrix} 0 & 1 & 0 & 0 \\ 0 & 0 & 0 & 0 \end{bmatrix}.$$

Then in the reduced form (3.3) only F is changed to

$$F = \begin{bmatrix} 1000 & 0 & 0 & 0 \\ 0 & 16178 & 10 & 1 \\ 0 & 1000 & 7 & 0 \\ 0 & 0 & 0 & 2 \end{bmatrix}$$

We obtain that $\alpha_0 = 1.9782$ is a singular value of Σ_F and

$$\begin{bmatrix} F_{22} & F_{23} & F_{24} \\ F_{32} & \Sigma_F & 0 \\ F_{42} & 0 & 0 \end{bmatrix}$$

has condition number 508.14 and norm of the inverse 0.5081. Choosing $\alpha = \alpha_0 + .0001$ we obtain $X = 16179$ and we have a new condition number 8194.0, which has increased, while the new norm of the inverse is 0.5055, which has decreased. This is not a typical situation, but it illustrates that minimizing the norm of the inverse may not always be the best choice.

EXAMPLE 4.5. In the last example we choose Let E, F, H be as in Example 4.4 and G as in Example 4.1. Then in the reduced form (3.3) only F is changed to

$$F = \begin{bmatrix} 1000 & 0 & 0 & 0 \\ 0 & -1 & 10 & 10 \\ 0 & 1 & 0 & -2 \\ 0 & 0 & 7.100 & 0 \end{bmatrix}$$

We obtain that $\alpha_0 = 2.0769$,

$$\begin{bmatrix} F_{22} & F_{23} & F_{24} \\ F_{32} & \Sigma_F & 0 \\ F_{42} & 0 & 0 \end{bmatrix}$$

has condition number 1739.4 and norm of the inverse 1.7394. In this case we obtain $X = \begin{bmatrix} -3.5715 \\ -0.5600 \end{bmatrix}$ and we have a new condition number 6.0797 and a norm of the inverse 0.4815.

We can see from these examples that the minimization of the norm of the inverse, if it is possible always makes the pencil regular index 1 and in most cases it also decreases significantly the condition number of the block to invert.

5. Conclusion. We have given explicit solutions for the problem of minimizing the norm of the inverse of a matrix, for which a submatrix can be assigned arbitrarily. The solution sets are characterized completely via the solution of Riccati inequalities. The obtained results are then applied to the regularization of descriptor control problems.

REFERENCES

[1] A. Bunse-Gerstner, V. Mehrmann, and N. Nichols, *Numerical methods for the regularization of descriptor systems by output feedback*, Tech. Rep. 987, Institute for Mathematics and its Applications, University of Minnesota, Minneapolis, Minnesota 55455, USA, 1992.

[2] ———, *Regularization of descriptor systems by derivative and proportional state feedback*, SIAM Journal Matrix Analysis and Applications, 13 (1992), pp. 46–67.

[3] L. Elsner, C. He, and V. Mehrmann, *Minimization of the norm, the norm of the inverse and the condition number of a matrix by completion*, Tech. Rep. 92-028, Sonderforschungsbereich 343, Diskrete Strukturen in der Mathematik, Universität Bielefeld, Fakultät für Mathematik.

[4] G. Golub and C. V. Loan, *Matrix Computations*, The Johns Hopkins Press, Baltimore, Maryland, second ed., 1989.

[5] D. Ouellette, *Schur complements and statistics*, Linear Algebra and its Applications, 36 (1981), pp. 187–295.

ON THE RUTISHAUSER'S APPROACH TO EIGENVALUE PROBLEMS *

L. FAYBUSOVICH[†]

Abstract. We consider the quotient-difference algorithm of H. Rutishauser from the point of view of the realization theory of rational functions. QR-like algorithms of the linear algebra can be thought of as discrete-time dynamical systems on certain classes of matrices. We show that these algorithms induce linear dynamical systems on various manifolds of rational functions. Applications to coding theory are considered.

1. Introduction. The work of H. Rutishauser proved to be fundamental in many areas of numerical analysis. In particular, it led to a broad class of algorithms for solving eigenvalue problems. H. Rutishauser developed his quotient-difference (QD) algorithm for the purpose of finding poles for various classes of meromorphic functions. He then realized that if the data related to QD-algorithm are organized in certain tridiagonal matrices, the QD-algorithm takes the form of what is now known as the LR-algorithm [9]. Later on various types of algorithms of this type have been developed. The rate of convergence of these algorithms (which we will call QR-like algorithms) depends on the choice of a variable shift strategy [9]. Although it is not difficult to incorporate these strategies into the original Rutishauser's approach, the questions of proving the convergence are still very complicated.

One of the best accounts of the original Rutishauser's work can be found in chapter 7 of [4] which appeared in 1974. It is interesting that in about the same time J. Moser [6] suggested to use the Rutishauser's idea (apparently without any knowledge of the Rutishauser's work) for the finding analytic solutions for a class of dynamical systems called Toda flows. A similar type of dynamical systems (continuous versions of the LR-algorithm) was discovered much earlier by H. Rutishauser [8]. See in this respect [10].

In the present paper we discuss H. Rutishauser's approach from the point of view of the realization theory of rational functions. Applications to coding theory are given.

2. QD-algorithm and LR-algorithm. Let $A \in GL(n, R)$ be an invertible n by n matrix. Suppose that all principal minors of A are nonzero. Then [9] A can be uniquely written in the form

$$A = LU. \tag{2.1}$$

* This work was partially supported by NSF through IMA grant.

† Department of Mathematics, University of Notre Dame, P.O. Box 398, Notre Dame, IN 46556.

Here U is an upper triangular matrix with units on the main diagonal (unit upper triangular matrix) and L is a lower triangular invertible matrix. Let $\alpha_0, \dots \alpha_n$ be given real numbers. Suppose that $A_0 \in gl(r, \mathbb{R})$ is such that $B_0 = A_0 - \alpha_0 I_r$ admits the factorization (2.1):

$$B_0 = L_0 U_0.$$

Set $A_1 = L_0^{-1} A_0 L_0 = U_0 L_0 + \alpha_0 I_n = U_0 A_0 U_0^{-1}$. Let we have already constructed $A_0, \dots A_i$ and $B_i = A_i - \alpha_i I_n$ admits the factorization (2.1):

$$B_i = L_i U_i.$$

Set $A_{i+1} = L_i^{-1} A_i L_i = (U_i L_i + \alpha_i I_n) = U_i A_i U_i^{-1}$. The described algorithm is known as the LR-algorithm with variable shifts $\alpha_0, \alpha_1, \dots$. The key observations [9] related to this algorithm are described as follows. First of all,

$$\begin{aligned} A_{i+1} &= L_i^{-1} A_i L_i = L_i^{-1} L_{i-1}^{-1} A_{i-1} L_{i-1} L_i = \dots \\ &= L_i^{-1} L_{i-1}^{-1} \dots L_0^{-1} A_0 L_0 L_1 \dots L_i \text{ or} \\ A_{i+1} &= (L_0 L_1 \dots L_i)^{-1} A_0 (L_0 L_1 \dots L_i). \end{aligned} \tag{2.2}$$

Similarly,

$$A_{i+1} = (U_i U_{i-1} \dots U_0) A_0 (U_i U_{i-1} \dots U_0)^{-1}. \tag{2.3}$$

Denote $L_0 L_1 \dots L_i$ by $\hat{L}_i$ and $U_i U_{i-1} \dots U_0$ by $\widehat{U}_i$. By (2.2) we have

$$\hat{L}_i (A_{i+1} - \alpha_{i+1} I_n) = (A_0 - \alpha_{i+1} I_n) \hat{L}_i.$$

But $A_{i+1} - \alpha_{i+1} I_n = L_{i+1} U_{i+1}$. Thus we can conclude that

$$\hat{L}_{i+1} U_{i+1} = (A_0 - \alpha_{i+1} I_n) \hat{L}_i. \tag{2.4}$$

If we know that

$$(A_0 - \alpha_0 I_n) \dots (A_0 - \alpha_i I_n) = \hat{L}_i \widehat{U}_i,$$

we derive from (2.4) that

$$(A_0 - \alpha_0 I_n) \dots (A_0 - \alpha_{i+1} I_n) = \hat{L}_{i+1} \widehat{U}_{i+1}. \tag{2.5}$$

Thus by induction we obtain that (2.5) is true for any i. For $A \in GL(n, \mathbb{R})$ introduce a rational function

$$f_A(z) = \langle e_1, (zI_n - A)^{-1} e_1 \rangle, \tag{2.6}$$

where $e_1, \dots e_n$ is the standard basis in $\mathbb{R}^n$ and

$$\langle x, y\rangle = x_1 y_1 + \dots + x_n y_n, \quad x, y \in \mathbb{R}^n.$$

PROPOSITION 2.1. *Let A_{i+1} be the i-th iterate of the LR-algorithm with shifts. Then*

$$(2.7)\quad f_{A_{i+1}}(z) = \frac{\langle e_1, (A_0-\alpha_0 I_n)\dots(A_0-\alpha_i I_n)(zI_n - A_0)^{-1}e_1\rangle}{\langle e_1, (A_0-\alpha_0 I_n)\dots(A_0-\alpha_i I_n)e_i\rangle},$$

$$(2.8)\quad f_{A_{i+1}}(z) = \frac{\langle e_1, (A_i-\alpha_i I_n)(zI_n - A_i)^{-1}e_1\rangle}{\langle e_1, (A_i-\alpha_i I_n)e_1\rangle}.$$

Proof. We prove (2.7). The proof of (2.8) is completely similar. By (2.3) we have

$$f_{A_{i+1}}(z) = \langle \widehat{U}_i^T e_1, (zI_n - A_0)^{-1}\widehat{U}_i^{-1}e_1\rangle.$$

Now $\widehat{U}_i^{-1}e_1 = e_1$, since $\widehat{U}_i$ is an upper triangular unit matrix. By (2.5) $\widehat{U}_i^T = \varphi_i(A_0)^T(\hat{L}_i^{-1})^T$, where $\varphi_i(x) = (x-\alpha_0)\dots(x-\alpha_i)$. Now $(L_i^{-1})^T$ is upper triangular. Thus $\widehat{U}_i^T e_i = \gamma\varphi_i(A_0)^T e_1$ for some nonzero γ. This yields

$$(2.9)\qquad f_{A_{i+1}}(z) = \gamma\langle e_1, \varphi_i(A_0)(zI_n - A_0)^{-1}e_1\rangle.$$

Let

$$(2.10)\qquad f_A(z) = \sum_{i=1}^{\infty}\frac{h_i(A)}{z^{i+1}}$$

be the expansion of f_A at the infinity. We clearly have

$$(2.11)\qquad h_i(A) = \langle e_1, A^i e_1\rangle.$$

In particular, $h_0(A) = 1$. Applying this observation to (2.9) we will have

$$(2.12)\qquad h_k(A_{i+1}) = \gamma\langle e_1, \varphi_i(A)A^k e_1\rangle.$$

In particular,

$$\gamma = \frac{1}{\langle e_1, \varphi_i(A)e_1\rangle}.$$

This completes the proof of (2.8). □

Let Rat($\mathbb{R}$) be the set of proper rational functions. It is clear that

$$\mathrm{Rat}(\mathbb{R}) = \cup_{n=1}^{\infty}\mathrm{Rat}(n, \mathbb{R}),$$

where $\mathrm{Rat}(n,\mathbb{R}) = \{\frac{p}{q} \in \mathrm{Rat}(\mathbb{R}) : p, q \in \mathbb{R}[x], \deg q = r, \deg p < r$ and p, q have no common divisors$\}$. In the control literature $\mathrm{Rat}(n,\mathbb{R})$ is usually called the manifold of proper rational functions of the McMillan degree r. It is convenient (as in [2]) to introduce the projectivization $P\,\mathrm{Rat}(\mathbb{R})$ of $\mathrm{Rat}(\mathbb{R})$ as follows: We identify two rational functions $f_1, f_2 \in \mathrm{Rat}(\mathbb{R})$ if $f_1 = \lambda f_2$ for some $\lambda \neq 0$. Denote by $\pi : \mathrm{Rat}(\mathbb{R}) \to P\,\mathrm{Rat}(\mathbb{R})$ the corresponding projection. Let $\pi(p/q) \in P\,\mathrm{Rat}(\mathbb{R})$, $\varphi \in \mathbb{R}[x]$. Let us define the map $M_\varphi : P\,\mathrm{Rat}(\mathbb{R}) \to P\,\mathrm{Rat}(\mathbb{R})$ as follows:

$$M_\varphi\left(\pi\left(\frac{p}{q}\right)\right) = \pi\left(\frac{\tilde{p}}{q}\right),$$

where $\tilde{p}$ is uniquely defined by the condition:

$$zp = \psi q + \tilde{p}, \quad \deg \tilde{p} < \deg q, \quad \psi \in \mathbb{R}[x].$$

Proposition 2.2. *The LR-algorithm with variable shifts induce discrete-time (time dependent) dynamical system D on $P\,\mathrm{Rat}(\mathbb{R})$:*

$$D\left(\pi\left(\frac{p}{q}\right), i\right) = M_{x-\alpha_i}\left(\pi\left(\frac{p}{q}\right)\right), i = 0, 1, \ldots .$$

Proof. By (2.8), (2.10)

$$\begin{aligned} h_K(A_{i+1}) &= \frac{\langle e_1, (A_i - \alpha_i I_n) A_i^K e_1\rangle}{\langle e_1, (A_i - \alpha_i I_n) e_1\rangle} = \\ &= \frac{h_{K+1}(A_i) - \alpha_i h_K(A_i)}{h_1(A_i) - \alpha_i h_0(A_i)}. \end{aligned} \tag{2.13}$$

Let $f_{A_i} = p/q$ and $(z - \alpha_i)p = \psi q + \tilde{p}$, $\deg \tilde{p} < r$. Since $\deg p < \deg q$, we clearly have: $\deg \psi \leq 1$. Thus $(z - \alpha_i)p/q$ differs from $\tilde{p}/q$ by a constant. If

$$p/q = \sum_{i=0}^{\infty} \frac{h_i}{z^{i+1}},$$

then

$$\frac{(z - \alpha_i)p}{q} = \sum_{i=0}^{\infty} \frac{h_{i+1} - \alpha_i h_i}{z^{i+1}} + h_0.$$

Hence

$$\frac{\tilde{p}}{q} = \sum_{i=0}^{\infty} \frac{h_{i+1} - \alpha_i h_i}{z^{i+1}}. \tag{2.14}$$

Comparing (2.13) with (2.14), we obtain

$$\pi\left(f_{A_{i+1}}\right)=\pi\left(M_{x-\alpha_i}\left(f_{A_i}\right)\right).$$

□

Let $\mathrm{Rat}_q(\mathbb{R}) = \{f = p/q \in \mathrm{Rat}(\mathbb{R}) : p, q$ have no common divisors$\}$.

COROLLARY 2.3. *Let $q(\alpha_i) \neq 0$, $i = 0, 1, \ldots$. The LR-algorithm with shifts $\alpha_0, \alpha_1, \ldots$ induces a dynamical system D on $P\,\mathrm{Rat}(\mathbb{R})$ which has $\mathrm{Rat}_q(\mathbb{R})$ as an invariant manifold.*

Remark. Let f_A be defined as in (2.6). The triple (e_1, A, e_1) is called a minimal realization of f_A if the vectors $e_1, Ae_1, \ldots A^{n-1}e_1$ and vectors $e_1, A^T e_1, \ldots (A^T)^{n-1}e_1$ form bases in $\mathbb{R}^n$.

PROPOSITION 2.4. *Suppose that (e_1, A_i, e_1)forms a minimal realization of f_{A_i}. Then (e_1, A_{i+1}, e_1) forms a minimal realization of $f_{A_{i+1}}$.*

Proof. We know that $A_{i+1} = U_i A_i U_i^{-1} = L_i^{-1} A_i L_i$, where U_i is unit upper triangular and L_i is lower triangular. We have $A_{i+1}^K e_1 = U_i A_i^K e_1$ for any K and $(A_{i+1}^T)^K e_1 = L_i^T (A_i^T)^K (L_i^{-T}) e_1$. Since $L_i^T e_1 = \gamma e_1$, γ is (1,1) entry of L_i, we easily obtain that the vectors $A_{i+1}^K e_1$, $(A_{i+1}^T)^K e_1$, $K = 0, 1, \ldots n-1$ form bases in $\mathbb{R}^n$. □

Let us introduce some notations. If L_i, U_i are matrices which appear in the description of the LR-algorithm, set

$$L_i = \|\ell_{st}^{(i)}\|, \quad U_i = \|u_{st}^{(i)}\|. \tag{2.15}$$

Earlier we introduced polynomials $\varphi_i(x) = (x-\alpha_0)(x-\alpha_1)\ldots(x-\alpha_i)$, $i = 0, 1, \ldots$. It is clear that $1, \varphi_0, \ldots \varphi_{i-1}$ form a basis in the space P_{i+1} of polynomials of degree less than $i+1$. We have $x = \varphi_0 + \alpha_0 1$, $x^2 = \varphi_1 + (\alpha_1 + \alpha_2)(\varphi_0 + \alpha_0) - \alpha_0\alpha_1$, etc. In general one can easily write down expressions

$$x^i = c_0^{(i)} 1 + c_1^{(i)} \varphi_0 + \ldots + c_{i-1}^{(i)} \varphi_{i-1}. \tag{2.16}$$

PROPOSITION 2.5. *In the notations introduced in (2.10), (2.15), (2.16)*

$$\begin{aligned} h_K(A_0) &= c_0^{(K)} + c_1^{(k)}\ell_{11}^{(0)} + \ldots + c_{i-1}^{(K)}\ell_{11}^{(0)}\ell_{11}^{(1)}\ldots\ell_{11}^{(K-1)}, \\ \kappa &= 0, 1, \ldots . \end{aligned} \tag{2.17}$$

Remark. In the case $\alpha_0 = \alpha_1 = \ldots = 0$, we have $x^i = \varphi_{i-1}(x)$. The expressions (2.17) are reduced to:

$$h_0(A_0) = 1, \quad h_K(A_0) = \ell_{11}^{(0)}\ell_{11}^{(1)}\ldots\ell_{11}^{(K-1)}, K = 1, 2, \ldots . \tag{2.18}$$

In particular,

$$\ell_{11}^{(K)} = \frac{h_{K+1}(A_0)}{h_K(A_0)}, K = 0, 1, \dots . \tag{2.19}$$

In general, equation (2.17) enable to find $\ell_{11}^{(K)}$ recursively. First, we find $\ell_{11}^{(0)}$ from the expression for $h_1(A_0)$. Then $\ell_{11}^{(1)}$ from the expression for $h_2(A_0)$ etc.

Proof. By (2.19), (2.16) $h_K(A_0) = \langle e_1, A_0^K e_1\rangle = c_0^{(K)} + c_1^{(K)}\langle e_1, \varphi_0(A_0)$ $e_1\rangle + \dots + c_{K-1}^{(K)}\langle e_1, \varphi_{K-1}(A_0)e_1\rangle$. By (2.5) we have $\varphi_i(A_0) = \hat{L}_i\widehat{U}_i$. Hence $\langle e_1, \varphi_i(A_0)e_1\rangle = \langle e_1, \hat{L}_i\widehat{U}_i e_1\rangle = \langle \hat{L}_i^T e_1, e_1\rangle = \langle L_i^T L_{i-1}^T \dots L_0^T e_1, e_1\rangle =$ $\ell_{11}^{(0)} \dots \ell_{11}^{(i)}\langle e_1, e_1\rangle$. Thus we arrive at (2.17). □

Let us now consider the matrices A of a special tridiagonal form:

$$A = \sum_{i=1}^{r} a_i E_{i,i} + \sum_{i=1}^{n-1} \mathcal{C}_i E_{i,i+1} + \sum_{i=1}^{n-1} E_{i+1,i}. \tag{2.20}$$

Here $\mathcal{C}_i$ are all nonzero and $E_{i,j}$ is the n by n matrix with 1 in the (i,j) entry and 0 in all other entries. If the principal minors of A are all nonzero, then it admits the LU-decomposition which has the special form:

$$L = \sum_{i=1}^{n-1} E_{i+1,i} \quad + \quad \sum_{i=1}^{n} \ell_{i,i} E_{i,i}, \tag{2.21}$$

$$U = \sum_{i=1}^{n} E_{i,i} \quad + \quad \sum_{i=1}^{n-1} u_{i,i+1} E_{i,i+1}. \tag{2.22}$$

We have

$$\begin{array}{rclrcl} a_i & = & \ell_{i,i} + u_{i-1,i}, & \mathcal{C}_i & = & \ell_{i,i} u_{i,i+1}, \\ i & = & 1, 2 \dots n-1, & a_n & = & \ell_{n,n} + u_{n-1,n}. \end{array} \tag{2.23}$$

Here $u_{0,1} = 0$. The matrix $\tilde{A} = UL$ will be again of the form (2.20) and

$$\tilde{a}_i = \ell_{i,i} + u_{i,i+1}, i = 1, 2, \dots n, \tag{2.24}$$

$$\tilde{\mathcal{C}}_i = \ell_{i+1,i+1} \cdot u_{i,i+1}, i = 1, 2, \dots n-1. \tag{2.25}$$

We can rewrite (2.23)–(2.25) as recursive relations for the LR-algorithm with shifts:

$$a_i^{(K)} - \alpha_K = \ell_{ii}^{(K)} + u_{i-1,i}^{(K)}, i = 1, 2, \dots n, \tag{2.26}$$

$$\mathcal{C}_i^{(K)} = \ell_{ii}^{(K)} u_{i,i+1}^{(K)}, i = 1, 2, \dots n-1, \tag{2.27}$$

$$a_i^{(K+1)} - \alpha_K = \ell_{ii}^{(K)} + u_{i,i+1}^{(K)}, i = 1, 2, \dots n, \tag{2.28}$$

$$\mathcal{C}_i^{(K+1)} = \ell_{i+1,i+1}^{(K)} \cdot u_{i,i+1}^{(K)}, i = 1, 2, \dots n-1. \tag{2.29}$$

Remark. α_K can be chosen as functions of $a_i^{(K)}, c_i^{(K)}$. For example, if we take $\alpha_K = a_n^{(K)}$, we arrive at the Rayleigh rule for the choice of variable shifts.

Suppose, first, that α_k does not depend on $a_i^{(K)}$, $c_i^{(K)}$ and the entries $\ell_{11}^{(K)}$, $K = 0, 1, 2, \ldots$, are known. Using the fact that $u_{0,1}^{(K)} = 0$ and (2.26), we can find $a_1^{(K)} = \alpha_K + \ell_{11}^{(K)}$, $K = 0, 1, 2, \ldots$. Then by (2.28) one can find $u_{12}^{(K)}, K = 0, 1, \ldots$. By (2.27) we then find $c_1^{(K)}$, $K = 0, 1, 2, \ldots$ and by (2.29) the entries $c_{22}^{(K)}$, $K = 0, 1, 2, \ldots$ can be determined. The general recursive scheme can be described as follows:

$$
\begin{aligned}
&(2.30) & a_i^{(K)} &= \alpha_K + \ell_{ii}^{(K)} + u_{i-1,i}^{(K)}, \\
&(2.31) & u_{i,i+1}^{(K)} &= a_i^{(K+1)} - \alpha_K - \ell_{ii}^{(K)}, \\
&(2.32) & c_i^{(K)} &= \ell_{ii}^{(K)} \cdot u_{i,i+1}^{(K)}, \\
&(2.33) & \ell_{i+1,i+1}^{(K)} &= c_i^{(K+1)} / u_{i,i+1}^{(K)}.
\end{aligned}
$$

It was the first version of the quotient-difference scheme (with $\alpha_K = 0$) suggested by H. Rutishauser. As initial conditions, the expressions (2.19) were used. Unfortunately this procedure turned out to be numerically unstable. We will show nevertheless that the scheme (2.30)–(2.33) can be used for decoding Goppa codes. The major version of the quotient-difference algorithm works as follows. The initial data $\ell_{ii}^{(0)}$, $u_{i,i+1}^{(0)}$, $i = 1, 2, \ldots n$, $u_{n,n+1}^{(0)} = 0$ should satisfy (2.26). We have the following recurrence relations:

$$
\begin{aligned}
&(2.34) & \ell_{ii}^{(K+1)} &+ u_{i-1,i}^{(K+1)} + \alpha_{K+1} = +\alpha_K + \ell_{i,i}^{(K)} + u_{i,i+1}^{(K)}, \\
&(2.35) & \ell_{ii}^{(K+1)} &\cdot u_{i,i+1}^{(K+1)} = \ell_{i+1,i+1}^{(K)} u_{i,i+1}^{(K)}.
\end{aligned}
$$

Relations (2.34), (2.35) (with $\alpha_K = 0$) form so-called "rhombus rules" of Rutishauser. These recurrence relations can be rewritten in the form:

$$
\begin{aligned}
&(2.36) & \ell_{ii}^{(K+1)} &= -(\alpha_K - \alpha_{K+1}) + \ell_{ii}^{(K)} + (u_{i,i+1}^{(K)} - u_{i-1,i}^{(K)}), \\
&(2.37) & u_{i,i+1}^{(K+1)} &= \frac{\ell_{i+1,i+1}^{(K+1)}, u_{i,i+1}^{(K)}}{\ell_{i,i}^{(K+1)}}, u_{0,1}^{(K)} = 0, K = 0, 1, \ldots .
\end{aligned}
$$

In this scheme α_K may depend on $a_i^{(K)}$, $c_i^{(K)}$. For example, $\alpha_K = \ell_{n,n}^{(K)} + u_{n-1,n}^{(K)}$ corresponds to the Rayleigh shift but the scheme (2.36), (2.37) should be modified in the following way:

$$
\begin{aligned}
&(2.38) & u_{i,i+1}^{(K)} &= c_i^{(K)} / (a_i^{(K)} - \alpha_K - u_{i-1,i}^{(K)}), \\
&(2.39) & \ell_{i,i}^{(K)} &= a_i^{(K)} - \alpha_K - u_{i-1,i}^{(K)},
\end{aligned}
$$

$$a_i^{(K+1)} = \ell_{ii}^{(K)} + u_{i,i+1}^{(K)} + \alpha_K, \tag{2.40}$$

$$c_i^{(K+1)} = \ell_{i+1,i+1}^{(K)} u_{i,i+1}^{(K)}. \tag{2.41}$$

Using (2.38), (2.39) and the condition $u_{0,1}^{(K)} = 0$, we first calculate $\ell_{ii}^{(K)}$, $u_{i,i+1}^{(K)}$, $i = 1, 2, \ldots r$. Then using (2.40), (2.41) one can find $a_i^{(K+1)}$, $c_i^{(K+1)}$.

3. Relations to the realization theory. We think of the LR-algorithm and the quotient-difference (QD) algorithm as a dynamical system D on the manifold of rational functions. This is possible provided there exists a one-to-one correspondence between certain rational functions and certain classes of matrices. This linkage is, of course, due to the realization theory of rational functions.

Suppose that $f \in \mathrm{Rat}(r, \mathbb{R})$ has the expansion at the infinity:

$$f(z) = \sum_{i=0}^{\infty} \frac{h_i}{z^{i+1}}. \tag{3.1}$$

We denote by P_{n-1} the vector space of polynomials with real coefficients of degree not greater than $n-1$. Introduce a bilinear symmetric form on P_{n-1}:

$$\langle\langle z^i, z^j \rangle\rangle = h_{i+j}, 0 \leq i, j \leq n-1. \tag{3.2}$$

Let $H_i = \|a_{st}\|$, $a_{st} = h_{s+t-2}$, $1 \leq s, t \leq i$ be Hankel matrices related to (3.1). Consider the basis $1, z, \ldots z^{n-1}$ in P_{n-1}. Suppose that $p_0, \ldots p_{n-1}$ are polynomials in P_{n-1} with the following properties: $\mathrm{span}(p_0, \ldots p_i) = \mathrm{span}(1, z, \ldots z^i)$, $i = 0, 1, \ldots n-1$; $p_i(z) = z^i + \ldots$; $\langle\langle p_i, p_j \rangle\rangle = 0$ if $i \neq j$. These polynomials are usually called Jacobi polynomials [4]. They exist if and only if $\det H_i \neq 0$, $i = 1, 2, \ldots n$ and admit the following well-known [4] description.

PROPOSITION 3.1.

$$p_i(z) = \frac{1}{\det H_i} \det \begin{bmatrix} h_0 & h_1 & \cdots & h_1 \\ h_1 & \cdots & \cdots & h_{i+1} \\ \vdots & & & \\ h_{i-1} & h_i & \cdots & h_{2i-1} \\ 1 & z & \cdots & z^i \end{bmatrix}. \tag{3.3}$$

Proof. It is clear from (3.3) that $p_i(z) = z^i + \ldots$. On the other hand, $\langle\langle z^j, p_i \rangle\rangle = 0, j < i$. □

As usual in the polynomial realization theory [2], introduce the shift operator $S : P_{n-1} \to P_{n-1}$ in the following way.

Let $f = p/q$, $\deg q = r$, $q(z) = z^n + a_1 z^{n-1} + \ldots + a_n$. We define $\mathcal{S}z^i = z^{i+1}, i = 0, 1, \ldots n - z$, $\mathcal{S}z^{n-1} = -a_1 z^{n-1} - \ldots - a_n$. In other words,

(3.4) $$\mathcal{S}z^i = z^{i+1} (\operatorname{mod} q).$$

PROPOSITION 3.2. *The shift operator $\mathcal{S}$ is symmetric relative to the scalar product (3.2).*

Proof. We should show that $\langle\langle \mathcal{S}z^i, z^j \rangle\rangle = \langle\langle z^i, \mathcal{S}z^j \rangle\rangle$ for any i, j. The only nontrivial case is where $i < n-1$, $j = n-1$. We have $\langle\langle \mathcal{S}z^i, z^{n-1} \rangle\rangle = h_{i+r}$, $\langle\langle z^i, \mathcal{S}z^{n-1} \rangle\rangle = \langle\langle z^i, -a_1 z^{n-1} - \ldots - a_r \rangle\rangle = -a_1 h_{n+i-1} - \ldots - a_n h_i$. The identities $h_{n+i} = -a_1 h_{n+i-1} - \ldots - a_n h_i$ follow from the observation:

(3.5) $$h_i = \frac{1}{2\pi\sqrt{-1}} \oint_C z^i f(z) dz,$$

where C is a contour in the complex plane which embraces all poles of f. Since it is also clear that

$$\oint_C z^i q(z) f(z) dz = 0, \quad i = 0, 1, \ldots,$$

we obtain the desired relations.

We have

$$\mathcal{S}p_i = p_{i+1} + c_i p_i + c_{i-1} p_{i-1} + \ldots + c_0, \qquad i = 0, 1, \ldots n-2.$$

Thus $\langle\langle \mathcal{S}p_i, p_j \rangle\rangle = c_j \langle\langle p_j, p_j \rangle\rangle$. But $\langle\langle \mathcal{S}p_i, p_j \rangle\rangle = \langle\langle p_i, \mathcal{S}p_j \rangle\rangle = 0$ if $j < i-1$. Hence we always have the following relations:

(3.6) $$\mathcal{S}p_i = p_{i+1} + a_{i+1} p_i + c_i p_{i-1}, i = 0, 1, \ldots n-2, p_{-1} = 0.$$

It remains to consider the case $i = n-1$, $\mathcal{S}p_{n-1} = c_{n-1} p_{n-1} + \ldots + c_0$. Thus $\langle\langle p_i, p_i \rangle\rangle c_i = \langle\langle \mathcal{S}p_{n-1}, p_i \rangle\rangle = \langle\langle p_{n-1}, \mathcal{S}p_i \rangle\rangle = 0$ if $i < n-2$. Thus

(3.7) $$\mathcal{S}p_{n-1} = a_r p_{n-1} + c_{n-1} p_{n-2}.$$

This means that in the basics $p_0, p_1, \ldots p_{n-1}$ the matrix of the operator $\mathcal{S}$ has exactly the form described in (2.20). Our final observation is:

(3.8) $$f(z) = \langle\langle 1, (zI_r - \mathcal{S})^{-1} 1 \rangle\rangle$$

It is sufficient to prove that

(3.9) $$h_i = \langle\langle 1, \mathcal{S}^i 1 \rangle\rangle, i = 0, 1, \ldots .$$

It is certainly true for $i = 0, 1, \ldots n$. Let $z^i = \psi(z)q(z) + r(z)$, $\psi, r \in \mathbb{R}[z]$, $\deg r < \deg q = n$. Since $q(\mathcal{S}) = 0$, we have $\mathcal{S}^i = r(\mathcal{S})$. By (3.5) we have

$$\begin{aligned} h_i &= \frac{1}{2\pi\sqrt{-1}} \oint_C z^i f(z) dz = \frac{1}{2\pi\sqrt{-1}} \int_C r(z) f(z) dz \\ &= \sum_{i=0}^{n-1} r_i h_i, \qquad \text{where } r(z) = \sum_{i=z}^{n-1} r_i z^i. \end{aligned}$$

On the other hand $\langle\langle \mathcal{S}^i 1, 1\rangle\rangle = \langle\langle r(S)1, 1\rangle\rangle = \sum_{i=1}^{n-1} r_i h_i$.

Consider now the matrix A defined as in (2.20). We want to prove that

$$f_A(z) = \langle\langle 1, (zI_n - \mathcal{S})^{-1} 1\rangle\rangle,$$

provided $h_0 = 1$ (see (2.6) for the definition of f_A). Indeed, if $\mathcal{S}^i 1 = c_0 p_0 + c_1 p_1 + \ldots + c_{i+1} p_{i+1}$, then $A^i e_1 = c_0 e_1 + \ldots + c_{i+1} e_{i+1}$. This simply follows from the fact that the operator A has the same matrix in the basis $e_1, \ldots e_n$ as the operator $\mathcal{S}$ has in the basis $p_0, \ldots p_{n-1}$. Thus $\langle e_1, A^i e_1\rangle = \langle e_1, c_0 e_1\rangle = \langle\langle \mathcal{S}^i \cdot 1, 1\rangle\rangle$. Thus we proved. □

PROPOSITION 3.3. *Let $f \in \mathrm{Rat}(n, \mathbb{R})$ and have the expansion (3.1) at the infinity. If $\det H_i \neq 0, i = 1, 2, \ldots, n$, $h_0 = 1$, then f admits a realization in the form (2.6) with A having the tridiagonal form (2.20).*

Inversely, suppose that $f \in \mathrm{Rat}(n, \mathbb{R})$ admits a realization in the form (2.6). Then all $\mathcal{C}_i \neq 0$. Otherwise, f admits a representation $f = p^1/q^1$ with $\deg q^1 < n$, since $f_A = f_{A^T}$. Contradicts the condition $f \in \mathrm{Rat}(n, \mathbb{R})$. This implies that $\mathrm{span}(e_1, \ldots e_i) = \mathrm{span}(e_1, Ae_1, \ldots, A^{i-1}e_1) = \mathrm{span}(e_1, A^T e_1, \ldots (A^T)^{i-1} e_1)$ for any i. In particular, e_1, $A^T e_1, \ldots (A^T)^{i-1}$ e_1 and $e_1, Ae_1, \ldots A^{i-1}$ e_1 are two bases of the same vector space. Thus the matrix $\Gamma_i = \| < A^{s-1}e_1, (A^T)^{t-1} e_1 \|$, $0 \leq s, t \leq i$ should be nondegenerate. But $\Gamma_i = H_i$. Suppose now that $f \in \mathrm{Rat}(n, \mathbb{R})$ admits two representations (2.6) with matrices A, A' having the tridiagonal form (2.20). As was observed earlier, $\mathcal{C}_i \neq 0$, $\mathcal{C}_i^1 \neq 0$, $i = 1, 2, \ldots n-1$. Since e_1, $Ae_1, \ldots A^{n-1}e_1$ and e_1, $A'e_1, \ldots (A^1)^{n-1} e_1$ form bases in $\mathbb{R}^n$, there exists an invertible n by n matrix C such that $CA^i e_1 = (A')^i e_1$, $i = 0, 1, \ldots, n-1$. Since $\mathrm{span}(e_1, Ae_1, \ldots A^{i-1} e_1) = \mathrm{span}(e_1, A'e_1, \ldots, (A')^{i-1} e_1) = \mathrm{span}(e_1, \ldots e_i)$, we have $C\,\mathrm{span}(e_1, \ldots e_i) = \mathrm{span}(e_1, \ldots e_i)$, $i = 1, 2, \ldots n$. Thus C is upper triangular. We obviously have $CAC^{-1} = A'$ and consequently $(C^{-1})^T A^T C^T = (A')^T$. Since $\langle C^T e_1, A^i e_1\rangle = \langle e_1, CA^i e_1\rangle = \langle e_1, (A')^i e_1\rangle = \langle e_1, A^i e_1\rangle$, $i = 0, 1, \ldots n-1$, we can conclude that $C^T e_1 = e_1$ (here we used $f_A = f_{A'}$). Applying the same reasoning as above to A^T, $(A')^T$ and C^T we obtain that C^T is also upper triangular. Thus C should be a diagonal matrix. But then $C = I_n$ since $CAC^{-1} = A'$, $Ce_1 = e_1$ and A, A' both have units on the subdiagonal. We proved the following theorem.

THEOREM 3.4. *Let $f \in \mathrm{Rat}(n, \mathbb{R})$ and have the expansion at the infinity (3.1). Then f admits a realization in the form (2.6) with A in the* *form (2.20) if and only if* $h_0 = 1$, $\det H_i \neq 0$, $i = 2, 3, \ldots, n$. *Such a representation is unique.*

Let us now return to the recurrence relations (3.6), (3.7). We obviously have: $\mathcal{C}_i\langle\langle p_{i-1}, p_{i-1}\rangle\rangle = \langle\langle \mathcal{S}p_i, p_{i-1}\rangle\rangle$, $i = 1, 2, \ldots n-1$. Now $\langle\langle \mathcal{S}p_i, p_{i-1}\rangle\rangle = \langle\langle p_i, \mathcal{S}p_{i-1}\rangle\rangle = \langle\langle p_i, zp_{i-1}\rangle\rangle = \langle\langle p_i, p_i\rangle\rangle$ since $p_i - zp_{i-1} \in P_{i-1}$. On the other hand, by (3.2)

$$\langle\langle p_i, p_i\rangle\rangle = \langle\langle p_i, z^i\rangle\rangle = \frac{\det H_{i+1}}{\det H_i}, i = 0, 1, \ldots, n-1,$$

where $\det H_0 = 1$. Thus

$$\mathcal{C}_i = \frac{\det H_{i+1} \det H_{i-1}}{\det H_i^2}, i = 1, 2, \ldots n-1. \tag{3.10}$$

Besides, $\mathcal{C}_1 \cdot \mathcal{C}_2 \ldots \mathcal{C}_i = \frac{\langle\langle p_1, p_1\rangle\rangle}{\langle\langle p_0, p_0\rangle\rangle} \cdot \frac{\langle\langle p_2, p_2\rangle\rangle}{\langle\langle p_1, p_1\rangle\rangle} \cdots \frac{\langle\langle p_i, p_i\rangle\rangle}{\langle\langle p_{i-1}, p_{i-1}\rangle\rangle} = \frac{\langle\langle p_i, p_i\rangle\rangle}{\langle\langle p_0, p_0\rangle\rangle} = \langle\langle p_i, p_i\rangle\rangle$, provided $h_0 = 1$.

Thus we obtain

$$\mathcal{C}_1 \ldots \mathcal{C}_i = \frac{\det H_{i+1}}{\det H_i}, \quad i = 1, 2, \ldots, n-1. \tag{3.11}$$

PROPOSITION 3.5. *Let* $h_0 = 1$, *then*

$$\operatorname{sign} H_{i+1} = \operatorname{sign} \mathcal{C}_1 \cdot \operatorname{sign} \mathcal{C}_2 \ldots \operatorname{sign} \mathcal{C}_i \operatorname{sign} H_i, \ldots i = 1, 2, \ldots, n-1.$$

We now want to get expressions for a_i in terms of Markov parameters h_i. Suppose that A has the form (2.20) and admits a LU-decomposition (2.21), (2.22). Set $\tilde{A} = UL$. If $f_A(z)$ has the expansion, then by (2.8)

$$f_{\tilde{A}}(z) = \sum_{i=0}^{\infty} \frac{h_{i+1}}{h_1 z^{i+1}}.$$

Using (2.27), (2.29), we obtain

$$\tilde{\mathcal{C}}_i = \frac{\ell_{i+1,i+1}}{\ell_{i,i}} \mathcal{C}_i, \quad i = 1, 2, \ldots, n-1, \quad \ell_{11} = h_1. \tag{3.12}$$

From (3.12) we have

$$\tilde{\mathcal{C}}_i \cdot \tilde{\mathcal{C}}_{i-1} \ldots \tilde{\mathcal{C}}_1 = \frac{\ell_{i+1,i+1}}{\ell_{1,1}} \mathcal{C}_i \cdot \mathcal{C}_{i-1} \ldots \mathcal{C}_1.$$

By (3.11) we finally obtain

$$\frac{\det \tilde{H}_{i+1}}{\det \tilde{H}_i} = \frac{\ell_{i+1,i+1}}{h_1} \frac{\det H_{i+1}}{\det H_i}. \tag{3.13}$$

Here $\widetilde{H}_i$ are the Hankel matrices for the function $f_{\tilde{A}}$. By (3.13) we have

$$\ell_{i,i} = \frac{\det H_{i-1} \det H_i^{(1)}}{\det H_i \det H_{i-1}^{(1)}}, \tag{3.14}$$

where

$$H_i^{(K)} = \begin{bmatrix} h_K & h_{K+1} & \cdots & h_{K+i-1} & \\ h_{K+1} & h_{K+2} & \cdots & h_{K+i} & \\ \vdots & & & & \\ h_{K+i-1} & h_{K+i} & \cdots & & h_{K+2i-2} \end{bmatrix},$$
$$\det H_0^{(K)} = 1.$$

By (2.27), (3.10)

$$u_{i,i+1} = \frac{c_i}{\ell_{i,i}} = \frac{\det H_{i+1} \det H_{i-1}^{(1)}}{\det H_i \det H_i^{(1)}}. \tag{3.15}$$

Finally by (2.26)

$$\begin{aligned} (3.16)\; a_i &= \ell_{ii} + u_{i-1,i} = \frac{\det H_{i-1} \det H_i^{(1)}}{\det H_i \det H_{i-1}^{(1)}} + \frac{\det H_i \det H_{i-2}^{(1)}}{\det H_{i-1} \det H_{i-1}^{(1)}}, \\ i &= 2, 3, \ldots n. \end{aligned}$$

Remark. We, of course, have similar expressions for $\tilde{\ell}_{ii}$, $\tilde{u}_{i,i+1}$. Now "rhombus rules" (2.34), (2.35) yield recurrence relations between Hankel determinants $H_i^{(K)}$. It is this relation that Henrici [4] uses to define the quotient-difference algorithm. Here we derive these relations starting from the quotient-difference scheme.

We can use relations (3.16), (3.10) to run the LR-algorithm with variable shifts as the dynamical system D described in Proposition 2.2 without using any matrix realizations. But involved calculations are, of course, too expensive.

4. The LR-algorithm and the HR-algorithm. The HR-algorithm was introduced and studied by A. Bunse-Gerstner [1] as a tool for solving certain eigenvalue problems. We begin with a brief description of this algorithm. Let D_K, $0 \leq K \leq n$ be the set of diagonal matrices with diagonal entries equal ± 1 such that exactly K of these entries are equal to -1. Given J_1, $J_2 \in D_K$, introduce the set $0(J_1, J_2) = \{0 \in GL(n, \mathbb{R}) : 0^T J_2 0 = J_1\}$. The iterations of the HR-algorithm are the sequence of pairs (A_i, J_i), where $A_i \in GL(n, \mathbb{R})$, $J_i \in D_K$ such that $A_i = 0_i R_i$, $0_i \in 0(J_{i+1}, J_i)$, R_i is upper triangular with positive entries on the main diagonal and $A_{i+1} = 0_i^{-1}A_i0_i$. Observe that J_{i+1} is determined by A_i, J_i as follows. Let $d_1, \ldots, d_n$ be the signs of principal minors of the matrix $A_i^T J_i A_i$. Then $J_{i+1} = \mathrm{diag}(\varepsilon_1, \ldots \varepsilon_n)$, where $\varepsilon_1\varepsilon_2 \ldots \varepsilon_K = d_K$, $K = 1, 2, \ldots n$. Let us denote by $\mathcal{S}(J)$ the set $\{A \in g\ell(n, \mathbb{R}) : A^T J = JA\}$, $J \in D_K$.

PROPOSITION 4.1. *If in the sequence of iterations, described above, $A_i \in \mathcal{S}(J_i)$, then $A_{i+1} \in \mathcal{S}(J_{i+1})$.*

Proof. Indeed $A_{i+1}^T J_{i+1} = 0_i^T A_i^T 0_i^{-1} J_{i+1} = 0_i^T A_i^T J_i(0_i^T) = 0_i^T J_i A_i 0_i = J_{i+1} 0_i^{-1} A_i 0_i = J_{i+1} A_{i+1}$. Here we used $0_i \in 0(J_{i+1}, J_i)$. For any $J \in D_k$ consider the map: $\mathcal{S}(J) \to \mathrm{Rat}(\mathbb{R})$ which is defined as follows.

$$f_A^J(z) = \langle e_1, J(zI_r - A)^{-1}e_1 \rangle.$$

□

PROPOSITION 4.2. *Suppose that $A \in \mathcal{S}(J)$, $A = 0 \cdot R$, $0 \in 0(\widetilde{J}, J)$, R is upper triangular with positive entries on the main diagonal. Then $\widetilde{A} = 0^{-1}A0 \in \mathcal{S}(\widetilde{J})$ and*

$$f_{\widetilde{A}}^{\widetilde{J}}(z) = \widetilde{\varepsilon}_1 \frac{\langle e_1, J(zI_r - A)^{-1}A^2 e_1 \rangle}{\langle e_1, JA^2 e_1 \rangle}, \tag{4.1}$$

where $\widetilde{J} = \mathrm{diag}(\widetilde{\varepsilon}_1, \ldots \widetilde{\varepsilon}_2)$.

Proof. We have

$$\begin{aligned} f_{\widetilde{A}}^{\widetilde{J}}(z) &= \langle e_1, \widetilde{J}0^{-1}(zI_r - A)^{-1}0e_1 \rangle \\ &= \langle e_1, 0^T J(zI_r - A)^{-1}0e_1 \rangle = \langle 0e_1, J(zI_r - A)^{-1}0e_1 \rangle \\ &= \langle AR^{-1}e_1, J(zI_r - A)^{-1}AR^{-1}e_1 \rangle = \\ &= \langle e_1, A^T J(zI_r - A)^{-1}Ae_1 \rangle = \\ &= \frac{\langle e_1, J(zI_r - A)^{-1}A^2 e_1 \rangle}{r_{11}^2}. \end{aligned}$$

Further,

$$\begin{aligned} r_{11}^2 &= \langle Re_1, Re_1 \rangle = \widetilde{\varepsilon}_1 \langle Re_1, \widetilde{J}Re_1 \rangle \\ &= \widetilde{\varepsilon}_1 \langle 0^{-1}Ae_1, \widetilde{J}0^{-1}Ae_1 \rangle = \widetilde{\varepsilon}_1 \langle Ae_1, (0^{-1})^T \widetilde{J}0^{-1}Ae_1 \rangle \\ &= \widetilde{\varepsilon}_1 \langle Ae_1, J00^{-1}Ae_1 \rangle = \widetilde{\varepsilon}_1 \langle e_1, A^T JAe_1 \rangle \\ &= \widetilde{\varepsilon}_1 \langle e_1, JA^2 e_1 \rangle. \end{aligned}$$

□

COROLLARY 4.3.

$$\pi(f_{\widetilde{A}}) = \pi(M_x^2(f_A)). \tag{4.2}$$

Comparing (4.2) with Proposition 2.2, we conclude that two steps of the LR-algorithm without shifts are equivalent to the one step of the HR-algorithm without shifts. This statement is reinforced by the following theorem.

THEOREM 4.4. *Let $T(J)$ be the set of tridiagonal matrices $\mathcal{S}$ in $\mathcal{S}(J)$ such that $\mathcal{S}_{i,i+1} > 0$, $i = 1, 2, \ldots n-1$. Let, further, $C(J) \subset \mathrm{Rat}(n, \mathbb{R})$ be the set of rational functions defined by conditions: sign* $\det H_k = \varepsilon_1 \varepsilon_2 \ldots \varepsilon_k$, *$k = 1, 2, \ldots n$, where $J = \mathrm{diag}(\varepsilon_1, \ldots \varepsilon_r)$. Then the map: $A \to \pi(f_A^J)$ defines an isomorphism of $\mathcal{S}(J)$ onto $\pi(C(J))$. For a proof see [3].*

PROPOSITION 4.5. *Under assumptions of Proposition 4.2, if $A \in T(J)$, then $\widetilde{A} \in T(\widetilde{J})$.*

Proof. If A is upper Hessenberg, then so is $\widetilde{A} = RAR^{-1}$. The result now follows from Proposition 10. □

Observe that the sets $\pi(C(J))$ are pairwise disjoint and their union forms an open dense subset in $P\,\mathrm{Rat}(n, \mathbb{R})$.

We can think of the HR-algorithm as a dynamical system on $P\,\mathrm{Rat}(n, \mathbb{R})$ described by (4.2). On each iteration the matrix interpretation can be obtained via Theorem 4.4.

5. A decoding algorithm for Goppa codes. Here we briefly describe one unexpected application of the quotient-difference algorithm. The details will be provided elsewhere. Suppose that $\mathcal{K}$ is a finite field, $g \in \mathcal{K}[x]$, $g(x) = (x - \alpha_1) \cdots (x - \alpha_{t-1})$, $\alpha_0, \ldots \alpha_{t-1} \in \mathcal{K}$. Given $\lambda_1, \cdots \lambda_r \in \mathcal{K}$, $g(\lambda_i) \neq 0$, $i = 1, 2, \ldots r$, consider the rational function

$$f(z) = \sum_{i=1}^{r} \frac{c_i}{g(\lambda_i)(z - \lambda_i)}. \tag{5.1}$$

Here, $c_i \in \mathcal{K}^*$, $i = 1, 2, \ldots r$.

We are interested in finding

$$\widetilde{f}(z) = \sum_{i=1}^{r} \frac{c_i}{z - \lambda_i}. \tag{5.2}$$

More precisely, the expansion of f at the infinity

$$f(z) = \sum_{i=0}^{\infty} \frac{h_i}{z^{i+1}}$$

is known. We want to recover $\widetilde{f}$. The Markov parameters h_i are called syndromes in the context of the decoding problem. The parameters λ_i correspond to error-location numbers and $c_1, \ldots c_r$ are error values [5].

We show here how the realization of

$$f(z) = \langle e_1, (zI_n - A)^{-1} e_1 \rangle$$

with A in the form (2.20) can be found. Here $e_1, \ldots e_r$ is the standard basis in $\mathcal{K}^r$ and

$$\langle x, y \rangle = x_1 y_1 + \cdots + x_r y_r, \quad x, y \in \mathcal{K}^r.$$

Observe that $\alpha_0, \ldots \alpha_{t-1}$ are supposed to be known. It is also known that $1 \le r \le \left[\frac{t}{2}\right]$. In our discussion of the LR-algorithm we never used the fact that our field is the field of real numbers (the only exception is Proposition 3.5). So all results related to LR-algorithm remain true for the field K. We need the following addition to Proposition 2.1.

PROPOSITION 5.1. *Let A_0, $A_1, \ldots A_i$ be iterates of the LR-algorithm with shifts $\alpha_0, \ldots \alpha_i$.*

If

$$f_{A_0}(z) = \sum_{i=1}^{r} \frac{c_i}{z - \lambda_i}, c_i \in \mathcal{K}^*, \lambda_i \in \mathcal{K},$$

then

$$f_{A_0}(z) = \gamma \sum_{i=1}^{r} \frac{g(\lambda_i) c_i}{z - \lambda_i},$$

where $g(x) = (x - \alpha_0) \ldots (x - \alpha_i), \gamma \in \mathcal{K}^*$.

Proof. Indeed, in this case $A_0 = V \operatorname{diag}(\lambda_1, \ldots \lambda_r) V^{-1}$ for some invertible matrix $V \in GL(r, \mathcal{K})$. The result now follows from (2.7). □

The idea of the application the LR-algorithm to decoding of Goppa codes is now clear. Take f defined by (5.1), perform t steps of the LR-algorithm with shifts $\alpha_0, \ldots \alpha_{t-1}$ and obtain (5.2) as the result. One can perform this procedure in terms of Markov parameters as in (2.13). There are two problems with this approach. First, we actually know only syndromes $h_0, \ldots h_{n-1}$, where n can be as small as $t+1$, which is, of course, sufficient to recover f provided $f \in \operatorname{Rat}(r, \mathcal{K})$, $r \le \left[\frac{t}{2}\right]$. But according to (2.13) we need to know $h_0, \ldots, h_{2t-1}$ to perform t steps of the LR-algorithm to obtain $\widetilde{h}_0, \ldots \widetilde{h}_{t-1}$ (Markov parameters of $\widetilde{f}$ defined by (5.2)). Second, we need to find c_i, λ_i out of $\widetilde{h}_0, \ldots \widetilde{h}_{t-1}$ which is a problem comparable in terms of complexity with the initial one. But in our situation we can use the original scheme of H. Rutishauser described in (2.30)–(2.33). The initial data $\ell_{11}^{(0)}, \dots \ell_{11}^{(t)}$ can be obtained from $h_0, \dots h_t$ using (2.17). Then we can find $a_i^{(t)}$, $i = 1, 2, \dots r$, $b_i^{(t)}$, $i = 1, 2, \dots r-1$. The number r will be defined by conditions $u_{r,r+1}^{(k)} = 0$ for $k = 0, 1, \dots t$.

Observe that the fact that these scheme is numerically unstable over reals is of no importance for the situation considered. Indeed, all calculations over finite field should and can be precise

6. Concluding remarks. In the present paper we considered the original approach of H. Rutishauser to the eigenvalue problems of linear algebra. The approach based on the realization theory enable to linearize various QR-like algorithms. Here we illustrated this by the example of the HR-algorithm. Finally, the version of the quotient-difference scheme was used to construct the decoding algorithm for Goppa codes. We strongly believe that the original Rutishauser's approach combined with ideas from the realization theory of rational and meromorphic functions can be very efficiently used in many situations which have not (and could not!) been addressed by H. Rutishauser himself.

REFERENCES

[1] A. Bunse-Gerstner. *An analysis of the HR-algorithm for computing the eigenvalues of a matrix.* Linear Algebra and its Applications, 35 (1981), pp. 155–173.

[2] R. Brockett and L. Faybusovich. *Toda flows, inverse spectral transform and realization theory.* Systems and Control Letters, 16 (1991), pp. 79–88.

[3] L. Faybusovich. *Hamiltonian and system-theoretic properties of the HR-algorithm.* (to appear in Journal of Mathematical Systems, Estimation and Control).

[4] P. Henrici. *Applied and Computational complex analysis.* Vol. 1, John Willey & Sons, 1974.

[5] R. Lidl and H. Niederreiter. *Introduction to finite fields and their applications.* Cambridge University Press, 1985.

[6] J. Moser. *Finitely many points on the line under the influence of an exponential potential – an integrable system.* In: J. Moser, Ed. Dynamical Systems, Theory and Applications, Lecture Notes in Physics, Vol. 38 (1975), pp. 467–497.

[7] H. Rutishauser. *Anwendunger des Quotienten – Differenzen – Algorithmus.* ZAMP, 5 (1954), pp. 496–508.

[8] H. Rutishauser. *Eir infinitesimales Analogon zum Quotienten – Differenzen – Algorithmus.* Arch. Math. 5 (1984), pp. 132–137.

[9] D. Watkins. *Fundamentals of matrix computations.* John Willey & Sons, 1991.

[10] D. Watkins and L. Elsner. *On Rutishauser's approach to self-similar flows.* SIAM journal on Matrix Analysis and Applications, 11 (1990), pp. 301–311.

THE BLOCK FORM OF LINEAR SYSTEMS OVER COMMUTATIVE RINGS WITH APPLICATIONS TO CONTROL

EDWARD W. KAMEN*

Abstract. This paper deals with a new approach to the study of linear time-invariant discrete-time systems whose coefficients belong to an arbitrary commutative ring. Such systems arise in the study of integer systems, systems depending on parameters, and multi-dimensional systems. The key idea is to consider the representation of systems over a ring in terms of a block-input/block-output form. By time-compressing the block representation, new results are derived on assignability by state feedback control including the construction of deadbeat controllers. A new type of state observer is then considered based on a block-output form for the update term in the state estimate. The results on state observers are combined with the time-compression approach to state feedback control to yield a new type of input/output regulator. In the last section of the paper the results are applied to the problem of state and output tracking of set points.

Key words. Systems over rings, reachability, observability, feedback control, observers, regulators, tracking

AMS(MOS) subject classifications. 93B05, 93B07, 93B25, 93C05, 93C45, 93C55, 93D15

1. Preliminaries. Let R be a commutative ring with multiplicative identity 1. Given positive integers m, n, p, consider the m-input p-output n-dimensional linear time-invariant discrete-time system over R given by the state model

$$x(k+1) = Ax(k) + Bu(k) \tag{1.1}$$

$$y(k) = Cx(k) \tag{1.2}$$

Here k is the discrete-time index, A, B, C are $n \times n$, $n \times m$, $p \times n$ matrices with entries in R, the input $u(k)$ is a m-element column vector with entries in R, the state $x(k)$ is a n-element column vector over R, and the output $y(k)$ is a p-element column vector over R.

Examples of classes of systems given by (1.1)-(1.2) include **integer systems** with $R = \mathbf{Z}$ = the ring of integers, **systems depending on N parameters** with R = a ring of real-valued continuous functions defined on a subset of N-dimensional Euclidean space, and **two-dimensional systems** with R = a convolution ring of real-valued functions defined on $\mathbf{Z}$ (see [1] for details).

Given the system (1.1)-(1.2) and given a positive integer q, we let Γ_q denote the $n \times mq$ **q-step reachability matrix** defined by

$$\Gamma_q = [B\ AB \dots A^{q-1}B] \tag{1.3}$$

* School of Electrical Engineering, Georgia Institute of Technology, Atlanta, GA 30332-0250.

Then the solution $x(k)$ of (1.1) at time $k = q$ starting from initial state $x(0)$ is

$$x(q) = A^q x(0) + \Gamma_q U_q \tag{1.4}$$

where U_q is the mq-element column vector

$$U_q = \begin{bmatrix} u(q-1) \\ u(q-2) \\ \vdots \\ u(0) \end{bmatrix}$$

The system (1.1)-(1.2) is said to be **reachable** (or **"ring" reachable**) if there is an integer $q \geq 1$ such that for any n-vector $\mathbf{x}$ over R, there is a U_q over R that "drives the system" from the zero initial state ($x(0) = 0$) to $x(q) = x$. It follows from (1.4) and is well known that the system (1.1)-(1.2) is reachable if and only if Γ_n has a right inverse over R. That is, there is a $mn \times n$ matrix Γ_n^r over R such that $\Gamma_n \Gamma_n^r = I_n$, where I_n is the $n \times n$ identity matrix.

Now let Ω_q denote the $pq \times n$ **q-step observability matrix** defined by

$$\Omega_q = \begin{bmatrix} C \\ CA \\ \vdots \\ CA^{q-1} \end{bmatrix} \tag{1.5}$$

If $u(k) = 0$ for $k = 0, 1, 2, \ldots q-1$, from (1.1), (1.2) we have that $Y_q = \Omega_q x(0)$ where Y_q is the pq-element column vector

$$Y_q = \begin{bmatrix} y(0) \\ y(1) \\ \vdots \\ y(q-1) \end{bmatrix}$$

The system (1.1)-(1.2) is **observable** if there is an integer $q \geq 1$ such that if $\Omega_q x = 0$ for some n-vector x over R, then x must be the zero vector. If R is a field, it is well known that observability is equivalent to left invertibility of Ω_n. When R is a ring, we say that the system (1.1)-(1.2) is **strongly observable** if Ω_n has a left inverse over R. Clearly, strong observability implies observability, but the converse is not true in general (unless R is a field).

If the system (1.1)-(1.2) is strongly observable, when $u(k) = 0$ for $k \geq 0$ the initial state $x(0)$ can be reconstructed from the block output vector Y_n

using the formula

$$x(0) = \Omega_n^\ell Y_n$$

where Ω_n^ℓ is a left inverse of Ω_n.

Denoting the system (1.1)-(1.2) by the triple (A, B, C), the **dual** of the system is the system over R given by (A^T, C^T, B^T) where superscript "T" denotes the transpose operation. It is obvious and well known that the given system (A, B, C) is strongly observable (resp. reachable) if and only if the dual (A^T, C^T, B^T) is reachable (resp. strongly observable).

2. The block form and time-compressed framework. Again consider the system (A, B, C) over the ring R given by the state model (1.1)-(1.2). Given an integer $q \geq 1$, iterating (1.1) $q-1$ times we have that

$$x(k+q) = A^q x(k) + \Gamma_q U_q(k) \tag{2.1}$$

where Γ_q is the q-step reachability matrix defined by (1.3) and $U_q(k)$ is the mq-element block input vector defined by

$$U_q^T(k) = [u^T(k+q-1) \;\; u^T(k+q-2) \ldots u^T(k)]$$

We can also express the output equation (1.2) in block form as follows. First, let E_q denote the $pq \times mq$ matrix

$$E_q = \begin{bmatrix} 0 & 0 & \cdots & 0 & 0 \\ 0 & 0 & \cdots & 0 & CB \\ 0 & 0 & \cdots & CB & CAB \\ \vdots & \vdots & & \vdots & \vdots \\ 0 & CB & \cdots & CA^{q-3}B & CA^{q-2}B \end{bmatrix}$$

Then iterating (1.1)-(1.2) $q-1$ times, we have the block form

$$Y_q(k) = \Omega_q x(k) + E_q U_q(k) \tag{2.2}$$

where $Y_q(k)$ is the pq-element block-output vector defined by

$$Y_q^T(k) = [y^T(k) \; y^T(k+1) \ldots y^T(k+q-1)]$$

and Ω_q is the q-step observability matrix defined by (1.5).

Equations (2.1) and (2.2) comprise the **block-input block-output state model** for the given system (1.1)-(1.2). Unfortunately, since $U_q(k+1)$ depends on $U_q(k)$ when $q > 1$ and $Y_q(k+1)$ depends on $Y_q(k)$ when $q > 1$, we cannot view (2.1)-(2.2) as the state model of a mq-input pq-output system. However, as shown below, by time scaling we can generate a mq-input pq-output representation.

We consider a time scaling by replacing k by kq in the time signals $U_q(k)$, $x(k)$, and $Y_q(k)$. If $q > 1$, then $U_q(kq)$, $x(kq)$, and $Y_q(kq)$ are time compressions of $U_q(k)$, $x(k)$, and $Y_q(k)$, respectively.

Replacing k by kq in (2.1) and (2.2), we obtain the system state model

$$x(kq+q) = A^q x(kq) + \Gamma_q U_q(kq) \tag{2.3}$$

$$Y_q(kq) = \Omega_q x(kq) + E_q U_q(kq) \tag{2.4}$$

When $q = 1$, (2.3)-(2.4) reduces to (2.1)-(2.2), but for $q > 1$, (2.3)-(2.4) is a time-compressed state model of the given system (1.1)-(1.2). In particular, one time step in (2.3)-(2.4) corresponds to q time steps in the original system model (1.1)-(1.2). Thus, we shall refer to (2.3)-(2.4) as the **q-scaled time-compressed model (TCM)**.

It should be noted that (2.3)-(2.4) corresponds to a generalization of multirate models [2,3,4] for linear time-invariant discrete-time systems defined over the reals. The time-compressed framework has also been extended to the study of linear time-varying discrete-time systems defined over the reals (see [5,6]).

Since $U_q(kq)$ is independent of $U_q((k+1)q)$ and $Y_q(kq)$ is independent of $Y_q((k+1)q)$, it is clear that (2.3)-(2.4) can be viewed as the state model of a mq-input pq-output n-dimensional linear time-invariant discrete-time system over the ring R. We will see that the given system (1.1)-(1.2) can be studied in terms of this framework. In particular, as a result of the increased number of inputs, it turns out that control is easier to carry out in the time-compressed framework.

3. State feedback control. Given the system (1.1)-(1.2) over the ring R, in this section we first consider the memoryless state feedback control

$$u(k) = -Fx(k), \ k \geq 0 \tag{3.1}$$

where F is a $m \times n$ matrix over R. (Here we are assuming that the time index begins at $k = 0$.) Inserting (3.1) into (1.1), we obtain the following state equation for the resulting closed-loop system

$$x(k+1) = (A - BF)x(k), k \geq 0 \tag{3.2}$$

The "characteristic polynomial" of the closed-loop system (3.2) is equal to

$$\det(zI_n - A + BF)$$

where "det" denotes the determinant, z is a symbol (or indeterminate), and I_n is the $n \times n$ identity matrix. Clearly, the characteristic polynomial can be expressed in the form

$$\det(zI_n - A + BF) = z^n + \sum_{i=0}^{n-1} a_i z^i \tag{3.3}$$

where the a_i are elements of R.

In the study of control for systems over a ring, a great deal of attention has been devoted to the question as to when the a_i in (3.3) can be chosen arbitrarily by choosing the feedback matrix F. This is the property of **coefficient assignability**; that is, the characteristic polynomial $\det(zI_n - A + BF)$ is coefficient assignable if for any $a_0, a_1, \ldots, a_{n-1}$ belonging to R, there is an F over R such that (3.3) holds.

If (3.3) holds, it follows from the Cayley-Hamilton theorem (for matrices over a commutative ring) that the state $x(k)$ of the closed-loop system (3.3) satisfies the relationship

$$x(k+n) = -\sum_{i=0}^{n-1} (a_i I_n) x(k+i), \; k \geq 0 \tag{3.4}$$

Since the a_i can be chosen arbitrarily, from (3.4) we see that coefficient assignability implies that the closed-loop system "dynamics" can be selected to be whatever one wants (in terms of the form (3.4)). In particular, if we can take $a_i = 0$ for all i, then (3.4) reduces to $x(k+n) = 0, \; k \geq 0$. In this case, the control $u(k) = -Fx(k)$ is called a **deadbeat control** since the state $x(k)$ is driven to zero.

The characteristic polynomial $\det(zI_n - A + BF)$ is said to be **pole assignable** if for any $p_1, p_2, \ldots, p_n$ belonging to R, there is an F over R such that

$$\det(zI_n - A + BF) = (z - p_1)(z - p_2)\ldots(z - p_n)$$

It is known that coefficient assignability implies pole assignability, which in turn implies reachability. However, the converse of these results is not valid in general (see [7]). If R is a field, it is well known that reachability, pole assignability, and coefficient assignability are all equivalent.

When R is a ring, it is also known that coefficient assignability may be achievable by considering a **dynamic state feedback** controller given by

$$u(k) = -Fx(k) + G\nu(k) \tag{3.5}$$

$$\nu(k+1) = H\nu(k) + Jx(k) \tag{3.6}$$

Here $\nu(k)$, which is an r-vector over R, is the state of the feedback controller, and F, G, H, J are $m \times n$, $m \times r$, $r \times r$, $r \times n$ matrices over R.

Combining (3.5)-(3.6) with (1.1)-(1.2), we obtain the following $(n+r)$ dimensional state equation for the resulting closed-loop system

$$\begin{bmatrix} x(k+1) \\ \nu(k+1) \end{bmatrix} = \begin{bmatrix} A-BF & BG \\ J & H \end{bmatrix} \begin{bmatrix} x(k) \\ \nu(k) \end{bmatrix} \tag{3.7}$$

The closed-loop characteristic polynomial is

$$\det \begin{bmatrix} zI_n - A + BF & -BG \\ -J & zI_r - H \end{bmatrix} \tag{3.8}$$

It is known [7] that if (1.1)-(1.2) is reachable, then there is a feedback controller of the form (3.5)-(3.6) such that the coefficients of (3.8) can be assigned arbitrarily. In particular, if the closed-loop characteristic polynomial is chosen to be z^{n+r}, then

$$\begin{bmatrix} x(k) \\ \nu(k) \end{bmatrix} = \begin{bmatrix} 0 \\ 0 \end{bmatrix}, \quad k \geq n+r$$

and thus we have deadbeat control.

The study of dynamic state feedback using controllers of the form (3.5)-(3.6) has turned out to be very challenging. In particular, the only constructions which currently exist for determining a dynamic state feedback controller of the form (3.5)-(3.6) **for a general class of commutative rings** yield a controller of dimension on the order of n^2 (see [7]). Obviously, this dimensionality can be rather excessive even for systems whose dimension n is not that large.

To make additional progress on the control of systems over a ring, in the next section we consider a new approach based on the time-compressed framework.

4. State feedback control in the time-compressed framework. In this section we illustrate the use of the q-scaled TCM (2.3)-(2.4) in control by considering state feedback in this framework.

Given a $mq \times n$ matrix F_q over R, we consider a type of state feedback control by setting

$$U_q(kq) = -F_q x(kq), \; k = 0, 1, 2, \ldots \tag{4.1}$$

Equation (4.1) specifies the control law in the time-compressed framework. To determine the corresponding control $u(k)$ for the original system model (1.1)-(1.2), we first write F_q in partitioned form as follows

$$F_q = \begin{bmatrix} F_{q1} \\ F_{q2} \\ \vdots \\ F_{qq} \end{bmatrix}$$

where the blocks F_{qi} are $m \times n$ matrices over R.

By definition of $U_q(k)$, from (4.1) we have that

$$\begin{bmatrix} u(kq+q-1) \\ u(kq+q-2) \\ \vdots \\ u(kq) \end{bmatrix} = -\begin{bmatrix} F_{q1} \\ F_{q2} \\ \vdots \\ F_{qq} \end{bmatrix} x(kq)$$

Hence

$$u(kq+q-i) = -F_{qi}x(kq), \; i = 1,2,\ldots,q, \; k = 0,1,2,\ldots$$

Replacing kq by k, we have

$$u(k+q-i) = -F_{qi}x(k), \; i = 1,2,\ldots,q, \; k = 0, \; q, \; 2q,\ldots \tag{4.2}$$

The control (4.2) can be expressed in the form

$$u(k) = -\sum_{i=0}^{q-1} G_i(k)x(k-i), \; k \geq q-1 \tag{4.3}$$

where the $G_i(k)$ are q-periodic $m \times n$ matrix functions of k with values in R; that is, $G_i(k+q) = G_i(k)$ for all $k \geq 0$. Note that the controller given by (4.3) has dimension equal to $q-1$.

From (4.3) we see that in the original time scale, the state feedback control (4.1) is **time varying** and **has dynamics** (i.e. it is not memoryless). Due to the time variance, the feedback control (4.3) differs from the time-invariant dynamic state feedback controller given by (3.5)-(3.6).

Inserting the control (4.3) into (1.1), we obtain the following closed-loop state equation in the original time scale

$$x(k+1) = [A - BG_0(k)]x(k) - \sum_{i=1}^{q-1} BG_i(k)x(k-i) \tag{4.4}$$

The closed-loop state equation in the kq time scale is given by

$$x(kq+q) = \left[A^q - \Gamma_q F_q\right] x(kq), \; k = 0,1,2,\ldots \tag{4.5}$$

The closed-loop equation (4.5) follows by inserting (4.1) into the q-scaled state equation (2.3).

The closed-loop system matrix $A^q - \Gamma_q F_q$ in (4.5) is said to be **assignable** if for any $n \times n$ matrix Q over R, there is a $mq \times n$ matrix F_q over R such that

$$A^q - \Gamma_q F_q = Q \tag{4.6}$$

If (4.6) holds, then the closed-loop system given by (4.5) satisfies the relationship

$$x(kq+q) = Qx(kq), \ k = 0,1,2,\ldots \tag{4.7}$$

From (4.7) we see that the closed-loop system's dynamics **on the kq time scale** are completely assignable if $A^q - \Gamma_q F_q$ is assignable.

We have the following result on assignability.

THEOREM 4.1. *The system matrix $A^q - \Gamma_q F_q$ is assignable for some $q \geq 1$ if and only if the given system (1.1)-(1.2) is reachable. In addition, if (1.1)-(1.2) is reachable so that Γ_n has a right inverse Γ_n^r over R, then for any $n \times n$ matrix Q over R, $A^n - \Gamma_n F_n = Q$ is satisfied with*

$$F_n = \Gamma_n^r(A^n - Q) \tag{4.8}$$

The proof of Theorem 4.1 is obvious, so we shall omit the details.

From Theorem 4.1 we see that reachability of the given system is equivalent to complete assignability on the kn time scale of the closed-loop system's dynamics. Of course this construction does not in general allow for the specification of arbitrary closed-loop dynamics on the original time scale. There is one very important case where assignability on the kq time scale does result in a corresponding result on the k time scale, and that is in deadbeat control.

To be precise, suppose that there is an F_q over R such that $A^q - \Gamma_q F_q = 0$. Then from (4.7), $x(kq+q) = 0$ for $k = 0,1,2,\ldots$, so we have deadbeat control on the kq time scale. But by (4.1), $U_q(kq) = 0$ for $k = 1,2,\ldots$, and thus by definition of $U_q(k)$, $u(k) = 0$ for $k = q, q+1, \ldots$ Then since $x(k+1) = Ax(k) + Bu(k)$, $x(k)$ must be zero for all $k \geq q$. Therefore, we also have deadbeat control on the k time scale.

Combining the above observations with Theorem 5.1, we have the following result.

THEOREM 4.2. *Suppose that the system (1.1)-(1.2) is reachable so that Γ_n has a right inverse Γ_n^r over R. Then the time-varying n-dimensional dynamic state feedback control*

$$U_n(kn) = \Gamma_n^r A^n x(kn), \ k = 0,1,2,\ldots \tag{4.9}$$

is a deadbeat control on the k time scale; i.e. with the control (4.9) the state $x(k)$ of the resulting closed-loop system is zero for all $k \geq n$.

By Theorem 4.2, reachability is a sufficient condition for the existence of a deadbeat control on the k time scale with the dimension of the controller (given by (4.3) with $q = n$) equal to $n-1$. This result is valid for systems over an arbitrary commutative ring R. Combining Theorems 4.1 and 4.2, we also have the following result.

THEOREM 4.3. *There is a deadbeat control of the form (4.1) if and only if*

$$A^q = \Gamma_q \Phi_q \tag{4.10}$$

for some $mq \times n$ matrix Φ_q over R.

Of course, if A has an inverse over R, the condition (4.10) is equivalent to requiring that Γ_q has a right inverse over R, which implies that the given system (1.1)-(1.2) is reachable.

5. State observers and input/output regulators. In the first part of this section, we first consider the generation of state observers by "dualizing" the results given in the previous section. Then the results on state feedback controllers and state observers are combined to generate a type of input/output regulator.

Given the block form (2.1)-(2.2) of the system (1.1)-(1.2), consider the state observer on the k time scale defined by

$$(5.1)\quad \widehat{x}(k+q)=A^q\widehat{x}(k)+L_q\left[Y_q(k)-\Omega_q\widehat{x}(k)-E_qU_q(k)\right]+\Gamma_qU_q(k),\ \ k\geq 0$$

In (5.1), $\widehat{x}(k)$ is the estimate of the state $x(k)$ at time k and L_q is the $n\times pq$ observer gain matrix defined over the ring R. Note that (5.1) differs from the standard form of a state observer in that the block output vector $Y_q(k)$ is used to update the estimate.

Defining the state estimation error

$$\widetilde{x}(k) = x(k) - \widetilde{x}(k)$$

and using (2.1), (2.2), and (5.1), we have that the error dynamics are given by

$$(5.2)\qquad \widetilde{x}(k+q) = \left[A^q - L_q\Omega_q\right]\widetilde{x}(k),\ \ k \geq 0.$$

The system matrix $A^q - L_q\Omega_q$ of the error equation (5.2) is said to be assignable if for any $n \times n$ matrix W over R, there is a $n \times pq$ matrix L_q over R such that

$$(5.3)\qquad A^q - L_q\Omega_q = W$$

If (5.3) holds, then the error equation (5.2) can be written in the form

$$(5.4)\qquad \widetilde{x}(k+1) = W\widetilde{x}(k),\ \ k \geq 0$$

Thus the error dynamics **on the k time scale** are completely assignable if $A^q - L_q\Omega_q$ is assignable.

Then "dualizing" Theorem 4.1, we have the following result.

THEOREM 5.1. *The system matrix $A^q - L_q\Omega_q$ is assignable for some $q \geq 1$ if and only if the given system (1.1)-(1.2) is strongly observable. If (1.1)-(1.2) is strongly observable, so that Ω_n has a left inverse Ω_n^ℓ over R, then for any $n \times n$ matrix W over R, $A^n - L_n\Omega_n = W$ satisfied with*

$$L_n = (A^n - W)\Omega_n^\ell$$

From Theorem 5.1 we see that strong observability of the given system is equivalent to complete assignability on the k time scale of the state estimation error dynamics. It is quite interesting that the assignability is on the k time scale; whereas in the control problem, the assignability is on the kn time scale.

We can generate a "deadbeat observer" by setting $W = 0$, so that (5.5) becomes

$$\widetilde{x}(k+1) = 0, \;\; k \geq 0$$

Thus, strong observability of the system (1.1)-(1.2) implies that we can exactly determine the system's state $x(k)$ after n steps. Dualizing Theorem 4.3, we have that there exists a deadbeat observer of the form (5.1) if and only if $A^q = P_q\Omega_q$ for some $n \times pq$ matrix P_q over R.

We can combine the controller results in the previous section with the above observer construction to yield an input/output regulator as follows.

Given the state feedback control (4.1), we replace $x(kq)$ by $\widehat{x}(kq)$ where $\widehat{x}(k)$ is the state estimate given by (5.1). This results in the following control law

$$U_q(kq) = -F_q\widehat{x}(kq), \;\; k = 0, 1, 2, \ldots \tag{5.5}$$

Note that in order to implement (5.5), it is only necessary to compute the estimate $\widehat{x}(kq)$ on the kq time scale. In other words, it is only necessary to evaluate (5.1) for $k = 0, q, 2q, \ldots$

Replacing k by kq in (5.2), we have that

$$\widetilde{x}(kq+q) = \big(A^q - L_q\Omega_q\big)\widetilde{x}(kq), \;\; k = 0, 1, 2, \ldots \tag{5.6}$$

and inserting (5.5) into (2.3), we obtain

$$x(kq+q) = A^q x(kq) - \Gamma_q F_q \widehat{x}(kq), \;\; k = 0, 1, 2, \ldots \tag{5.7}$$

But $\widehat{x}(kq) = x(kq) - \widetilde{x}(kq)$, and thus from (5.7),

$$x(kq+q) = \big[A^q - \Gamma_q F_q\big] x(kq) + \Gamma_q F_q \widetilde{x}(kq), \;\; k = 0, 1, 2, \ldots \tag{5.8}$$

Then combining (5.6) and (5.8), we have that the closed-loop system resulting from the control (5.5) is given by

$$\begin{bmatrix} x(kq+q) \\ \widetilde{x}(kq+q) \end{bmatrix} = \begin{bmatrix} A^q - \Gamma_q F_q & \Gamma_q F_q \\ 0 & A^q - L_q\Omega_q \end{bmatrix} \begin{bmatrix} x(kq) \\ \widetilde{x}(kq) \end{bmatrix}, \;\; k=0, 1, 2, \ldots \tag{5.9}$$

Equation (5.9) is the state model of the closed-loop system in the kq time scale.

Now if the system (1.1)-(1.2) is reachable and strongly observable, then by Theorems 4.1 and 5.1, for any $n \times n$ matrices Q and W over R, there exist matrices F_n and L_n over R such that the system matrix in (5.9) is equal to

$$\begin{bmatrix} Q & \Gamma_n F_n \\ 0 & W \end{bmatrix} \tag{5.10}$$

In other words, the system matrix of the closed-loop system on the kn time scale can be assigned up to the extent indicated by (5.10). In particular, if Q and W are chosen to be zero, then from (5.9) we have that

$$x(kn+n) = \Gamma_n F_n \widetilde{x}(kn), \; k = 0, 1, 2, \ldots$$

$$\widetilde{x}(kn+n) = 0, \; k = 0, 1, 2, \ldots$$

Combining these two equations, we have that

$$x(kn) = 0, k = 2, 3, 4, \ldots$$

Thus we have deadbeat control on the kn time scale. But since

$$U_n(kn) = -F_n \widehat{x}(kn), \; k = 0, 1, 2, \ldots \tag{5.11}$$

and $\widehat{x}(kn) = x(kn) = 0$ for $k = 2, 3, 4, \ldots$ and since $x(k+1) = Ax(k) + Bu(k)$, it must be true that $x(k) = 0, k = 2n, 2n+1, 2n+2, \ldots$. Thus, the control (5.11) with the observer (5.1) is a "deadbeat input/output regulator" on the k time scale.

Combining the above results with the results in the previous section, we have the following conclusions.

THEOREM 5.2. *A sufficient condition for the existence of a deadbeat input/output regulator (given by (5.5) with the observer (5.1) is that the system (1.1)-(1.2) be reachable and strongly observable. A necessary and sufficient condition for the existence of a deadbeat input/output regulator given by (5.5) and (5.1) is that there exist an integer $q \geq 1$ and matrices Φ_q and Γ_q over R such that $A^q = \Gamma_q \Phi_q = P_q \Omega_q$.*

6. Application to tracking. In this section we utilize the results derived in the previous sections to solve a type of tracking problem, called **set-point control**. We begin with set-point control of the state $x(k)$ of the system (1.1)-(1.2), and then we consider set-point control of the output $y(k)$.

Given an n-vector x_d over R, in state set-point control the objective is to determine a control input $u(k)$ which forces the state $x(k)$ to be equal to x_d, where x_d is the (desired) set-point. Since the system is given by $x(k+1) = Ax(k) + Bu(k)$, the existence of a set-point control $u(k)$ requires that there exists an m-vector u_∞ over R such that

$$x_d = Ax_d + Bu_\infty$$

or

$$(A - I)x_d = -Bu_\infty \tag{6.1}$$

Suppose that there exists a u_∞ over R satisfying (6.1), and define

$$\begin{aligned}\overline{x}(k) &= x(k) - x_d \\ \overline{u}(k) &= u(k) - u_\infty\end{aligned}$$

Then

$$\begin{aligned}\overline{x}(k+1) &= x(k+1) - x_d = Ax(k) + Bu(k) - x_d \\ \overline{x}(k+1) &= A[\overline{x}(k) + x_d] + B\big[\overline{u}(k) + u_\infty\big] - x_d \\ \overline{x}(k+1) &= A\overline{x}(k) + B\overline{u}(k) + (A - I)x_d + Bu_\infty\end{aligned}$$

Thus using (6.1), we have that

$$\overline{x}(k+1) + A\overline{x}(k) + B\overline{u}(k) \tag{6.2}$$

We can write (6.2) in the block-input form

$$\overline{x}(k+1) = A^q\overline{x}(k) + \Gamma_q\overline{U}_q(k) \tag{6.3}$$

where

$$\overline{U}_q(k) = [\overline{u}^T(k+q-1)\overline{u}^T(k+q-2)\ldots\overline{u}^T(k)]^T$$

Time compressing (6.3) gives

$$\overline{x}(kq+q) = A^q\overline{x}(kq) + \Gamma_q\overline{U}_q(kq)$$

Now suppose that for some $q \geq 1$, $A^q = \Gamma_q\Phi_q$ for some $mq \times n$ matrix Φ_q over R. Then by Theorem 4.3, the state feedback control

$$\overline{U}_q(kq) = \Phi_q A^q\overline{x}(kq) \tag{6.4}$$

results in a deadbeat control on the k time scale. That is, $\overline{x}(k) = 0$ for $k = q,\ q+1,\ q+2, \ldots$. Hence, $x(k) = x_d$ for $k = q, q+1, q+2, \ldots$, and so the control (6.4) results in set-point tracking. In addition, by definition of $\overline{u}(k)$, the control (6.4) can be rewritten in the form

$$U_q(kq) = \Phi_q A^q[x(kq) - x_d] + \left[u_\infty^T u_\infty^T \cdots u_\infty^T\right]^T \tag{6.5}$$

We then have the following result.

THEOREM 6.1. *Given the desired set-point x_d, there is a deadbeat set-point tracking control of the form (6.4) or (6.5) if and only if there is* *an m-vector* u_∞ *over* R *such that* $(A-I)x_d = -Bu_\infty$ *and there exist an integer* $q \geq 1$ *and a matrix* Φ_q *over* R *such that* $A^q = \Gamma_q \Phi_q$.

Now given an m-vector y_d over R, in output set-point control we want to find a control that will force the output $y(k)$ to be equal to y_d, where y_d is the (desired) set-point. The existence of such a control requires that there exist an m-vector u_∞ over R and a n-vector x_∞ over R such that

$$\begin{aligned} x_\infty &= Ax_\infty + Bu_\infty \\ y_d &= Cx_\infty \end{aligned}$$

or

$$\begin{bmatrix} A-I & -B \\ C & 0 \end{bmatrix} \begin{bmatrix} x_\infty \\ u_\infty \end{bmatrix} = \begin{bmatrix} 0 \\ y_d \end{bmatrix} \tag{6.6}$$

Sufficient conditions for (6.6) to be satisfied are that $A-I$ is invertible over R and there exists an m-vector u_∞ over R such that

$$C(I-A)^{-1}Bu_\infty = y_d \tag{6.7}$$

It is interesting to note that when R is a field, invertibility of $C(I-A)^{-1}B$ is a necessary and sufficient condition for tracking an arbitrary set-point, and that this condition is equivalent to requiring that the system has no zeros at $z=1$.

Using the above results, we have the following conclusion.

THEOREM 6.2. *Suppose that* $A-I$ *is invertible over* R *and there is a* u_∞ *over* R *satisfying (6.7). Also suppose that* $A^q = \Gamma_q\Phi_q$ *for some* Φ_q *over* R. *Then the control (6.5) with* $x_d = (I-A)^{-1}Bu_\infty$ *results in deadbeat output set-point tracking; that is,* $y(k) = y_d$ *for* $k = q,\ q+1, \ldots$. *In addition,* $x(k) = x_d$ *for* $k = q, q+1, \ldots$.

The set-point tracking controllers given above can be generalized to the case when the state $x(kq)$ must be estimated by using an observer of the form (5.1). If the given system is strongly observable, and the conditions given above are satisfied, we can construct deadbeat trackers. The details are not pursued here.

REFERENCES

[1] E.W. KAMEN, *Linear systems over rings: from R.E. Kalman to the present*, in Mathematical System Theory-The Influence of R.E. Kalman, A.C. Antoulas (Ed.), Springer-Verlag, Berlin (1991), pp. 311–324.

[2] P. ALBERTOS, *Block multirate input-output model for sampled control systems*, IEEE Trans. on Automatic Control, 35 (1990), pp. 1085–1088.

[3] P. ALBERTOS AND R. ORTEGA, *On generalized predictive control: two alternative formulations*, Automatica, 25 (1989), pp. 753–755.

[4] R. LOZANO, *Robust adaptive regulation without persistent excitation*, IEEE Trans. on Automatic Control, 34 (1989), pp. 1260–1267.

[5] E.W. Kamen, *Study of linear time-varying discrete-time systems in terms of time-compressed models*, Proceedings of the 31st IEEE Decision and Control Conference, Tucson, AZ (1992), pp. 3070–3075.

[6] E.W. Kamen, *Regulation of time-varying discrete-time systems based on a time-compressed formulation*, Proceedings of the International Symposium on Implicit and Nonlinear Systems, Arlington, TX (1992), pp. 267–274.

[7] J.W. Brewer, J.W. Bunce, and F.S. Vanvleck, *Linear systems over commutative rings*, Marcel Dekker, New York, 1986.

DIFFEOMORPHISMS BETWEEN SETS OF LINEAR SYSTEMS

R. OBER* AND P. A. FUHRMANN†

Abstract. Diffeomorphisms are given between different subsets of linear systems of fixed McMillan degree. The sets considered are the set of all systems of fixed McMillan degree, the subset of stable systems, the subset of bounded real systems, the subset of positive real systems, the subset of stable systems with Hankel singular values bounded by one. State space techniques are used in the proofs.

1. Introduction. In such areas as system identification, time series analysis or controller design by parameter optimization, often a nonlinear search has to be performed over a specific set of linear systems of fixed McMillan degree. It is therefore important to have knowledge of the structure of such classes of systems. This motivated Brockett [2] to study the topology of the set of single-input single-output linear systems of fixed McMillan degree n. He proved that this set has $n+1$ connected components. Glover [9] showed that in the multivariable situation there is however only one connected component. The same result was established by Hanzon [12] and independently by Ober [19] for the sub-class of (asymptotically) stable systems of McMillan degree n. Results of this type suggested that there might be a close connection between these two classes of systems. Helmke [15] then showed that these two classes are homeomorphic, and Hanzon [13] used a different approach to show that they are in fact diffeomorphic.

Using different types of balanced realizations, Ober [19], [21] and Ober and McFarlane [20] derived canonical forms for several classes of linear systems of fixed McMillan degree: stable, positive real, bounded real, minimum phase and systems without constraints. An interesting aspect of these canonical forms is that they have a remarkably similar structure. This gave further evidence that there should be a strong relationship between these classes of systems. In Ober [20] it was moreover shown that all these sets of systems have identical numbers of connected components.

In this paper we will establish new diffeomorphisms between sets of systems. The diffeomorphism between the set of all systems of fixed McMillan degree and the subset of stable systems is motivated by a map that was studied in much detail in Fuhrmann-Ober [8] and by some state space formulae in Glover-McFarlane [10]. The other diffeomorphisms studied in this paper are in fact adaptations of this map. These maps also explain to a

* Center for Engineering Mathematics, Programs in Mathematical Sciences, The University of Texas at Dallas, Richardson, Texas 75083 – 06688 , USA.

† Earl Katz Family Chair in Algebraic System Theory.

Partially supported by the Israeli Academy of Sciences, Department of Mathematics, Ben-Gurion University of the Negev, Beer Sheva, Israel.

great extent why the canonical forms for the different classes of systems in Ober [21] have such a similar structure.

For single-input single-output systems it was shown in Ober [21] that each minimal, or stable, or bounded real or positive real system can be parametrized by a set of *standard parameters*:

$$
\begin{array}{ll}
\sigma_1 > \ldots > \sigma_j > \ldots > \sigma_k > 0 & \\
n_1, \ldots, n_j, \ldots, n_k & n_j \in \mathcal{N},\ \sum_{j=1}^{k} n_j = n; \\
s_1, \ldots, s_j, \ldots, s_k & s_j = \pm 1,\quad 1 \le j \le k; \\
b_1, \alpha(1)_1, \ldots, \alpha(1)_j, \ldots, \alpha(1)_{n_1-1}, & b_1 > 0,\ \alpha(1)_j > 0,\ 1 \le j \le n_1 - 1; \\
\vdots & \\
b_i, \alpha(i)_1, \ldots, \alpha(i)_j, \ldots, \alpha(i)_{n_i-1}, & b_i > 0,\ \alpha(i)_j > 0,\ 1 \le j \le n_i - 1; \\
\vdots & \\
b_k, \alpha(k)_1, \ldots, \alpha(k)_j, \ldots, \alpha(k)_{n_k-1}, & b_k > 0,\ \alpha(k)_j > 0,\ 1 \le j \le n_k - 1; \\
d & d \in \Re
\end{array}
$$

In particular each system in one of the classes of systems has a unique representation, a canonical form, in terms of following 'standard system'.

The *standard system* (A, b, c, d) is then given by

1. $b = (\underbrace{b_1, 0, \ldots, 0}_{n_1}, \ldots, \underbrace{b_j, 0, \ldots, 0}_{n_j}, \ldots, \underbrace{b_k, 0, \ldots, 0}_{n_k})^T$,
2. $c = (\underbrace{s_1 b_1, 0, \ldots, 0}_{n_1}, \ldots, \underbrace{s_j b_j, 0, \ldots, 0}_{n_j}, \ldots, \underbrace{s_k b_k, 0, \ldots, 0}_{n_k})$,
3. For $A =: (A_{ij})_{1 \le i,j \le k}$ we have
   a) block diagonal entries $A_{jj},\ 1 \le j \le k$:

$$
A_{jj} = \begin{pmatrix}
a_{jj} & \alpha(j)_1 & & & & \\
-\alpha(j)_1 & 0 & \alpha(j)_2 & & & \\
& -\alpha(j)_2 & 0 & \cdot & & 0 \\
& & \ddots & \ddots & \ddots & \\
& 0 & & \cdot & 0 & \alpha(j)_{n_j-1} \\
& & & & -\alpha(j)_{n_j-1} & 0
\end{pmatrix}
$$

with a_{jj} a function of b_j, σ_j and d.

   b) off diagonal blocks A_{ij}, $1 \le i, j \le k$, $i \ne j$:

$$
A_{ij} = \begin{pmatrix}
a_{ij} & 0 & \ldots & 0 \\
0 & 0 & \ldots & 0 \\
\vdots & \vdots & & \vdots \\
0 & 0 & \ldots & 0
\end{pmatrix}
$$

with a_{ij} a function of $b_i, b_j, s_i, s_j, \sigma_i, \sigma_j$ and d.

For the case of stable systems we have the following canonical form (Ober [18]). We call a minimal system stable if all its poles are in the open left half plane.

THEOREM 1.1. *The following two statements are equivalent:*

(i) $g(s)$ is the transfer function of a stable minimal system over $\Re$ of McMillan degree n.

(ii) $g(s)$ has a standard n-dimensional realization (A, b, c, d) given by a standard set of parameters such that

$$a_{ij} = \frac{-b_i b_j}{s_i s_j \sigma_i + \sigma_j}.$$

Moreover, the map which assigns to each stable minimal system the realization in (ii) is a canonical form.

The canonical form presented in the previous theorem is in the form of a Lyapunov balanced realization (Moore [17]).

DEFINITION 1.2. *A stable minimal system (A, B, C, D) is called* Lyapunov balanced *if*

$$A\Sigma + \Sigma A^* = -BB^*,$$

$$A^*\Sigma + \Sigma A = -C^*C,$$

with $\Sigma = diag(\sigma_1 I_{n_1}, \sigma_2 I_{n_2}, \ldots, \sigma_k I_{n_k})$, $\sigma_1 > \sigma_2 > \cdots > \sigma_k > 0$. The matrix Σ is called the Lyapunov grammian *of the system.*

The canonical form quoted in the previous theorem is Lyapunov balanced with Lyapunov grammian $\Sigma = diag(\sigma_1 I_{n_1}, \sigma_2 I_{n_2}, \ldots, \sigma_k I_{n_k})$. Another interesting property of the canonical form is that it is sign-symmetric. Indeed if

$$S = diag(s_1 \hat{I}_{n_1}, \ldots, s_j \hat{I}_{n_j}, \ldots, s_k \hat{I}_{n_k}),$$

where $\hat{I}_{n_j} = diag(+1, -1, +1, \ldots, (-1)^{n_j+1}) \in \Re^{n_j \times n_j}$, $j = 1, \ldots, k$, then

$$A = SA^T S, \quad c^T = Sb.$$

It should also be noted that the Cauchy index of a system is given by $trace(S)$ (see Anderson [1]).

In Section 2 we introduce a map, the so-called L-characteristic, that maps not necessarily stable minimal systems to stable minimal systems of the same McMillan degree. It will be shown that this characteristic map is in fact a bijection between the set of minimal systems of fixed McMillan degree and its subset of all stable minimal systems. This map will also be analyzed from the point of view LQG-balanced realizations and Lyapunov realizations. Sections 3-5 contain the analogous analyzes for bounded real systems, positive real systems and antistable systems. Finally in Section 6 it is shown that the bijections are in fact diffeomorphisms.

Both authors would like to thank the Institute for Technomathematik and in particular Prof. D. Praetzel-Wolters at the University of Kaiserslautern, Germany for their hospitality during the preparation of this manuscript. Paul Fuhrmann would like to gratefully acknowledge the support of the DFG (Deutsche Forschungsgemeinschaft) during his stay in Kaiserslautern.

2. Minimal systems. The aim of this section is to establish a bijection between the set of minimal systems of dimension n and the set of all stable minimal systems of dimension n. Note that we mean by a stable system a system whose poles are all in the open left half plane. This bijection which we call the L-characteristic map is in fact a map that occurred implicitly in the work by Glover and McFarlane [10] and was analyzed from an operator theoretic point of view in Fuhrmann and Ober [8].

To simplify presentation we introduce the notation,

$$A_L := A - B(I + D^*D)^{-1}D^*C,$$

for a given linear system (A, B, C, D). Note that $A_L = A$ if $D = 0$.

DEFINITION 2.1. *Let* $\left(\begin{array}{c|c} A & B \\ \hline C & D \end{array}\right)$ *be a minimal system. Let Y be the stabilizing solution of the Riccati equation*

$$0 = A_L^*Y + YA_L - YB(I + D^*D)^{-1}B^*Y + C^*(I + DD^*)^{-1}C,$$

i.e. $A_L - B(I + D^*D)^{-1}B^*Y$ is stable, and let Z be the stabilizing solution to the Riccati equation

$$0 = A_LZ + ZA_L^* - ZC^*(I + DD^*)^{-1}CZ + B(I + D^*D)^{-1}B^*,$$

i.e. $A_L - ZC^*(I + DD^*)^{-1}C$ is stable.

Then the system

$$\chi_L\left(\left(\begin{array}{c|c} A & B \\ \hline C & D \end{array}\right)\right) := \left(\begin{array}{c|c} A_L - B(I + D^*D)^{-1}B^*Y & B(I + D^*D)^{-1/2} \\ \hline (I + DD^*)^{-1/2}C(I + ZY) & D \end{array}\right)$$

is called the L-characteristic *of the system* $\left(\begin{array}{c|c} A & B \\ \hline C & D \end{array}\right)$.

The following relationships are due to Bucy [3]. Since the reference is difficult to find we give a short proof.

LEMMA 2.2 (BUCY RELATIONSHIPS). *Let* $\left(\begin{array}{c|c} A & B \\ \hline C & D \end{array}\right)$ *be a minimal system. Let Y be a solution of the Riccati equation*

$$0 = A_L^*Y + YA_L - YB(I + D^*D)^{-1}B^*Y + C^*(I + DD^*)^{-1}C,$$

and let Z be a solution of the Riccati equation

$$0 = A_LZ + ZA_L^* - ZC^*(I + DD^*)^{-1}CZ + B(I + D^*D)^{-1}B^*,$$

then

$$[I+ZY](A_L - B(I+D^*D)^{-1}B^*Y) = (A_L - ZC^*(I+DD^*)^{-1}C)[I+ZY].$$

Proof. Consider the two Riccati equations,

$$0 = A_L^*Y + YA_L - YB(I+D^*D)^{-1}B^*Y + C^*(I+DD^*)^{-1}C,$$

and

$$0 = A_LZ + ZA_L^* - ZC^*(I+DD^*)^{-1}CZ + B(I+D^*D)^{-1}B^*.$$

Multiplying the first equation on the left by Z and the second equation on the right by Y, equating both equations and adding A_L to both sides we obtain

$$A_L + ZA_L^*Y + ZYA_L - ZYB(I+D^*D)^{-1}B^*Y + ZC^*(I+DD^*)^{-1}C$$

$$= A_L + A_LZY + ZA_L^*Y - ZC^*(I+DD^*)^{-1}CZY + B(I+D^*D)^{-1}B^*Y.$$

Canceling the term ZA_L^*Y from either side and collecting terms, we obtain

$$[I+ZY](A_L - B(I+D^*D)^{-1}B^*Y) = (A_L - ZC^*(I+DD^*)^{-1}C)[I+ZY].$$

□

As a consequence of the Bucy relationships we can rewrite the L-characteristic of a system as follows,

$$\chi_L\left(\left(\begin{array}{c|c} A & B \\ \hline C & D \end{array}\right)\right) :=$$

$$\left(\begin{array}{c|c} [I+ZY]^{-1}(A_L - ZC^*(I+DD^*)^{-1}C)[I+ZY]) & B(I+D^*D)^{-1/2} \\ \hline (I+DD^*)^{-1/2}C(I+ZY) & D \end{array}\right).$$

The following Lemma shows that the L-characteristic map maps a system with no stability assumptions to a stable system of the same McMillan degree.

LEMMA 2.3. *The L-characteristic of a minimal system is stable and minimal. The L-characteristics of two equivalent systems are equivalent.*

Proof. Since Y is the stabilizing solution of the Riccati equation the matrix $A_L - B(I+D^*D)^{-1}B^*Y$ is stable by definition. It is easily seen that the characteristic system is reachable. The observability of the system follows by using the representation of the characteristic in which the A matrix is written in the form resulting from the Bucy relations.

Let $(A,B,C,D) \in L_n^{p,m}$. If Z is the stabilizing solution to the Riccati equation,

$$A_LZ + ZA_L^* - Z(C^*(I+DD^*)^{-1}CZ + B(I+D^*D)^{-1}B^* = 0,$$

then TZT^* is the stabilizing solution to this Riccati equation for the system $(TAT^{-1}, TB, CT^{-1}, D)$, where T is non-singular. Similarly, if Y is the stabilizing solution to

$$A_L^* Y + YA_L - YB(I + D^*D)^{-1}B^*Y + C^*(I + DD^*)^{-1}C = 0,$$

then $T^{-*}YT^{-1}$ is the stabilizing solution to this Riccati equation for the system $(TAT^{-1}, TB, CT^{-1}, D)$. Using this fact it is easily seen that the L-characteristic of two equivalent systems are equivalent. □

The main theorem of this section will show that the L-characteristic map is in fact a bijection between the set of n-dimensional minimal systems and the set of stable n-dimensional minimal systems. We denote by $L_n^{p,m}$ the set of all minimal n-dimensional systems, with m-dimensional input and p-dimensional output space. The subset of continuous-time stable systems is denoted by $C_n^{p,m}$. Recall that we mean by a stable system what is often referred to as an asymptotically stable system, i.e. all the eigenvalues of the A matrix are in the open left half plane.

In the next definition we are going to define the so-called inverse L-characteristic map $I\chi_L : C_n^{p,m} \to L_n^{p,m}$. We will show that this map is in fact the inverse of the L-characteristic map χ_L.

DEFINITION 2.4. *Let $(\mathcal{A}, \mathcal{B}, \mathcal{C}, \mathcal{D}) \in C_n^{p,m}$ and let P and Q be the solutions to the Lyapunov equations*

$$\mathcal{A}P + P\mathcal{A}^* = -\mathcal{B}\mathcal{B}^*, \quad \mathcal{A}^*Q + Q\mathcal{A} = -\mathcal{C}^*\mathcal{C}.$$

Then

$$I\chi_L\left(\left(\begin{array}{c|c} \mathcal{A} & \mathcal{B} \\ \hline \mathcal{C} & \mathcal{D} \end{array}\right)\right) := \left(\begin{array}{c|c} \mathcal{A}+\mathcal{B}(\mathcal{B}^*Q+\mathcal{D}^*\mathcal{C})(I+PQ)^{-1} & \mathcal{B}(I+\mathcal{D}^*\mathcal{D})^{1/2} \\ \hline (I+\mathcal{D}\mathcal{D}^*)^{1/2}\mathcal{C}(I+PQ)^{-1} & \mathcal{D} \end{array}\right)$$

is called the inverse L-characteristic system.

In order to be able to analyze the inverse L-characteristic map we need the following Lemma.

LEMMA 2.5. *Let $(\mathcal{A}, \mathcal{B}, \mathcal{C}, \mathcal{D}) \in C_n^{p,m}$ and let P and Q be such that*

$$\mathcal{A}P + P\mathcal{A}^* = -\mathcal{B}\mathcal{B}^*,$$

$$\mathcal{A}^*Q + Q\mathcal{A} = -\mathcal{C}^*\mathcal{C}.$$

Then

$$[\mathcal{A}+\mathcal{B}(\mathcal{B}^*Q+\mathcal{D}^*\mathcal{C})(I+PQ)^{-1}][I+PQ] = [I+PQ][\mathcal{A}+(I+PQ)^{-1}(P\mathcal{C}^*+\mathcal{B}\mathcal{D}^*)\mathcal{C}]$$

Proof. We have

$$[\mathcal{A} + \mathcal{B}(\mathcal{B}^*Q + \mathcal{D}^*\mathcal{C})(I + PQ)^{-1}][I + PQ] = \mathcal{A}(I + PQ) + \mathcal{B}(\mathcal{B}^*Q + \mathcal{D}^*\mathcal{C})$$

$$= \mathcal{A} + \mathcal{A}PQ + \mathcal{B}\mathcal{B}^*Q + \mathcal{B}\mathcal{D}^*\mathcal{C} = \mathcal{A} + (\mathcal{A}P + \mathcal{B}\mathcal{B}^*)Q + \mathcal{B}\mathcal{D}^*\mathcal{C}$$

$$= \mathcal{A} + (-P\mathcal{A}^*)Q + \mathcal{B}\mathcal{D}^*\mathcal{C} = \mathcal{A} - P(-Q\mathcal{A} - \mathcal{C}^*\mathcal{C}) + \mathcal{B}\mathcal{D}^*\mathcal{C}$$

$$= \mathcal{A} + PQ\mathcal{A} + (P\mathcal{C}^* + \mathcal{B}\mathcal{D}^*)\mathcal{C}$$

$$= [I + PQ][\mathcal{A} + (I + PQ)^{-1}(P\mathcal{C}^* + \mathcal{B}\mathcal{D}^*)\mathcal{C}].$$

□

The following Lemma shows that the inverse L-characteristic system is minimal.

LEMMA 2.6. *We have that*

$$I\chi_L(C_n^{p,m}) \subseteq L_n^{p,m}.$$

Proof. Let $(\mathcal{A}, \mathcal{B}, \mathcal{C}, \mathcal{D}) \in C_n^{p,m}$. That $I\chi_L((\mathcal{A}, \mathcal{B}, \mathcal{C}, \mathcal{D}))$ is reachable follows immediately from the given representation. The observability follows in a similar way after rewriting $\mathcal{A} + \mathcal{B}(\mathcal{B}^*Q + \mathcal{D}^*\mathcal{C})(I + PQ)^{-1}$ as $(I + PQ)(\mathcal{A} + (I + PQ)^{-1}(P\mathcal{C}^* + \mathcal{B}\mathcal{D}^*)\mathcal{C})(I + PQ)^{-1}$, using Lemma 2.5. □

The following proposition shows that the characteristic map is injective. We need the following Lemma that shows how the solutions of the Riccati equations of a minimal system are related to the solutions of Lyapunov equations of its L-characteristic system.

LEMMA 2.7. *Let* $\left(\begin{array}{c|c} A & B \\ \hline C & D \end{array}\right)$ *be a minimal system and let*

$$\left(\begin{array}{c|c} \mathcal{A} & \mathcal{B} \\ \hline \mathcal{C} & \mathcal{D} \end{array}\right) := \left(\begin{array}{c|c} A_L - B(I + D^*D)^{-1}B^*Y & B(I + D^*D)^{-1/2} \\ \hline (I + DD^*)^{-1/2}C(I + ZY) & D \end{array}\right)$$

be its L-characteristic system, with Y *and* Z *the solutions to the respective Riccati equations. Then the Lyapunov equations*

$$\mathcal{A}P + P\mathcal{A}^* = -\mathcal{B}\mathcal{B}^*$$

$$\mathcal{A}^*Q + Q\mathcal{A} = -\mathcal{C}^*\mathcal{C}$$

have solutions given by

$$P = (I + ZY)^{-1}Z = Z(I + YZ)^{-1}$$

$$Q = Y(I + ZY) = (I + YZ)Y.$$

Proof. We want to show that with $P = (I + ZY)^{-1}Z = Z(I + YZ)^{-1}$ we have,

$$\mathcal{A}P + P\mathcal{A}^* = -\mathcal{B}\mathcal{B}^*.$$

To do this consider

$$(I+ZY)[\mathcal{A}P+P\mathcal{A}^*](I+YZ)$$

$$= (I+ZY)[(A_L - B(I+D^*D)^{-1}B^*Y)P + P(A_L - B(I+D^*D)^{-1}B^*Y)^*] \\ (I+YZ)$$

$$= (I+ZY)(A_L - B(I+D^*D)^{-1}B^*Y)Z + Z(A_L - B(I+D^*D)^{-1}B^*Y)^* \\ (I+YZ)$$

$$= A_LZ + ZA_L^* + Z(YA_L + A_L^*Y)Z$$

$$-2ZYB(I+D^*D)^{-1}B^*YZ - B(I+D^*D)^{-1}B^*YZ - ZYB(I+D^*D)^{-1}B^*,$$

using the two Riccati equations this gives,

$$= ZC^*(I+DD^*)^{-1}CZ - B(I+D^*D)^{-1}B^*$$

$$+Z[YB(I+D^*D)^{-1}B^*Y - C^*(I+DD^*)^{-1}C)]Z$$

$$-2ZYB(I+D^*D)^{-1}B^*YZ - B(I+D^*D)^{-1}B^*YZ - ZYB(I+D^*D)^{-1}B^*$$

$$= -(I+ZY)B(I+D^*D)^{-1}B^*(I+YZ)$$

$$= -(I+ZY)\mathcal{B}\mathcal{B}^*(I+YZ),$$

which shows the claim.

Now with $Q = Y + YZY$, we have

$$\mathcal{A}^*Q + Q\mathcal{A} = \mathcal{A}^*Y(I+ZY) + (I+YZ)Y\mathcal{A}$$

$$= (A_L^*Y - YB(I+D^*D)^{-1}B^*Y)(I+ZY) + (I+YZ)(YA_L - YB \\ (I+D^*D)^{-1}B^*Y),$$

using the Riccati equation, we have

$$= (-YA_L - C^*(I+DD^*)^{-1}C)(I+ZY) + (I+YZ)(-A_L^*Y - C^*(I+DD^*)^{-1}C)$$

$$= -C^*(I+DD^*)^{-1}C(I+ZY) - (I+YZ)C^*(I+DD^*)^{-1}C$$

$$-A_L^*Y - YA_L - Y(A_LZ + ZA_L^*)Y$$

$$= -C^*(I+DD^*)^{-1}C(I+ZY) - (I+YZ)C^*(I+DD^*)^{-1}C$$

$$-YB(I+D^*D)^{-1}B^*Y + C^*(I+DD^*)^{-1}C - Y(ZC^*(I+DD^*)^{-1}CZ)$$

$$-B(I+D^*D)^{-1}B^*)Y$$

$$= -(I+YZ)C^*(I+DD^*)^{-1}C(I+ZY)$$

$$= -\mathcal{C}^*\mathcal{C}.$$

□

We can now prove the proposition.

PROPOSITION 2.8. *The characteristic map* χ_L *is injective. More precisely,* $I\chi_L \cdot \chi_L$ *is the identity map on* $L_n^{p,m}$*.*

Proof. Let $(A,B,C,D) \in L_n^{p,m}$ and let $(\mathcal{A},\mathcal{B},\mathcal{C},\mathcal{D}) \in C_n^{p,m}$ be its L-characteristic, i.e.

$$\left(\begin{array}{c|c} \mathcal{A} & \mathcal{B} \\ \hline \mathcal{C} & \mathcal{D} \end{array}\right) = \left(\begin{array}{c|c} A_L - B(I+D^*D)^{-1}B^*Y) & B(I+D^*D)^{-1/2} \\ \hline (I+DD^*)^{-1/2}C(I+ZY) & D \end{array}\right),$$

where Y and Z are the stabilizing solutions to the respective Riccati equations. We know by Lemma 2.7 that the solutions to the Lyapunov equations

$$\mathcal{A}P + P\mathcal{A}^* = -\mathcal{B}\mathcal{B}^*, \quad \mathcal{A}^*Q + Q\mathcal{A} = -\mathcal{C}^*\mathcal{C}$$

are given by

$$P = (I+ZY)^{-1}Z = Z(I+YZ)^{-1},$$

$$Q = Y(I+ZY) = (I+YZ)Y.$$

Hence we can see that $PQ = ZY$. Now apply $I\chi_L$ to $(\mathcal{A},\mathcal{B},\mathcal{C},\mathcal{D})$ and set $(A_1,B_1,C_1,D_1) := I\chi_L((\mathcal{A},\mathcal{B},\mathcal{C},\mathcal{D}))$, then $D_1 = D$ and

$$B_1 = \mathcal{B}(I+\mathcal{D}^*\mathcal{D})^{1/2} = B(I+D^*D)^{-1/2}(I+D^*D)^{1/2} = B,$$

$$C_1 = (I+\mathcal{D}\mathcal{D}^*)^{1/2}\mathcal{C}(I+PQ)^{-1}$$

$$= (I+DD^*)^{1/2}(I+DD^*)^{-1/2}C(I+ZY)(I+ZY)^{-1} = C,$$

$$A_1 = \mathcal{A} + \mathcal{B}\mathcal{B}^*Q(I+PQ)^{-1} + \mathcal{B}\mathcal{D}^*\mathcal{C}(I+PQ)^{-1}$$

$$= A - B(I+D^*D)^{-1}(D^*C+B^*Y) + B(I+D^*D)^{-1}B^*Y(I+ZY)(I+ZY)^{-1}$$

$$+B(I+D^*D)^{-1}D^*C(I+ZY)(I+ZY)^{-1}$$

$$= A,$$

i.e. $I\chi_L \cdot \chi_L((A,B,C,D)) = (A,B,C,D)$ for $(A,B,C,D) \in L_n^{p,m}$. □

We now need to prove that χ_L is in fact surjective, or that $\chi_L \cdot I\chi_L$ is the identity map on $C_n^{p,m}$. To do this we need the following Lemma.

LEMMA 2.9. *Let $(\mathcal{A},\mathcal{B},\mathcal{C},\mathcal{D}) \in C_n^{p,m}$. Let P, Q be the positive definite solutions to the Lyapunov equations*

$$\mathcal{A}P + P\mathcal{A}^* + \mathcal{B}\mathcal{B}^* = 0, \quad \mathcal{A}^*Q + Q\mathcal{A} + \mathcal{C}^*\mathcal{C} = 0.$$

Let

$$\left(\begin{array}{c|c} A & B \\ \hline C & D \end{array}\right) := I\chi_L\left(\left(\begin{array}{c|c} \mathcal{A} & \mathcal{B} \\ \hline \mathcal{C} & \mathcal{D} \end{array}\right)\right).$$

Then

$$0 = A_L^*Y + YA_L - YB(I+D^*D)^{-1}B^*Y + C^*(I+DD^*)^{-1}C,$$

$$0 = A_LZ + ZA_L^* - ZC^*(I+DD^*)^{-1}CZ + B(I+D^*D)^{-1}B^*,$$

with

$$Y = Q(I+PQ)^{-1} = (I+QP)^{-1}Q,$$

$$Z = P(I+QP).$$

Moreover, Y and Z are the stabilizing solutions to the Riccati equations.

Proof. First note that

$$A_L = A - B(I+D^*D)^{-1}D^*C$$

$$= \mathcal{A} + \mathcal{B}(\mathcal{B}^*Q + \mathcal{D}^*\mathcal{C})(I+PQ)^{-1}$$

$$-\mathcal{B}(I+\mathcal{D}^*\mathcal{D})^{1/2}(I+\mathcal{D}^*\mathcal{D})^{-1}\mathcal{D}^*(I+\mathcal{D}\mathcal{D}^*)^{1/2}\mathcal{C}(I+PQ)^{-1}$$

$$= \mathcal{A} + \mathcal{B}\mathcal{B}^*Q(I+PQ)^{-1}.$$

Since,

$$(I+QP)[A_L^*Y + YA_L - YB(I+D^*D)^{-1}B^*Y + C^*(I+DD^*)^{-1}C](I+PQ)$$

$$= (I + QP)[(\mathcal{A} + \mathcal{B}\mathcal{B}^* Q(I + PQ)^{-1})^* Q(I + PQ)^{-1}$$

$$+(I + QP)^{-1} Q(\mathcal{A} + \mathcal{B}\mathcal{B}^* Q(I + PQ)^{-1})$$

$$-(I + QP)^{-1} Q\mathcal{B}(I + \mathcal{D}^*\mathcal{D})^{1/2}(I + \mathcal{D}^*\mathcal{D})^{-1}(I + \mathcal{D}^*\mathcal{D})^{1/2}\mathcal{B}^* Q(I + PQ)^{-1}$$

$$+(I+PQ)^{-*}\mathcal{C}^*(I+\mathcal{D}\mathcal{D}^*)^{1/2}(I+\mathcal{D}\mathcal{D}^*)^{-1}(I+\mathcal{D}\mathcal{D}^*)^{1/2}\mathcal{C}(I+PQ)^{-1}](I+PQ)$$

$$= (I + QP)\mathcal{A}^* Q + Q\mathcal{B}\mathcal{B}^* Q + Q\mathcal{A}(I + PQ) + Q\mathcal{B}\mathcal{B}^* Q - Q\mathcal{B}\mathcal{B}^* Q + \mathcal{C}^*\mathcal{C}$$

$$= \mathcal{A}^* Q + Q\mathcal{A} + \mathcal{C}^*\mathcal{C} + Q(P\mathcal{A}^* + \mathcal{A}P + \mathcal{B}\mathcal{B}^*)Q$$

$$= 0,$$

we have verified the first identity. Now with $Z = P(I + QP)$ we have

$$A_L Z + Z A_L^* - ZC^*(I + DD^*)^{-1} CZ + B(I + D^* D)^{-1} B^*$$

$$= (\mathcal{A}+\mathcal{B}\mathcal{B}^* Q(I+PQ)^{-1})(I+PQ)P + P(I+QP)(\mathcal{A}+\mathcal{B}\mathcal{B}^* Q(I+PQ)^{-1})^*$$

$$-P(I + QP)(I + PQ)^{-*}\mathcal{C}(I + PQ)^{-1}(I + PQ)P$$

$$+\mathcal{B}(I + \mathcal{D}^*\mathcal{D})^{1/2}(I + \mathcal{D}^*\mathcal{D})^{-1}(I + \mathcal{D}^*\mathcal{D})^{1/2}\mathcal{B}^*$$

$$= \mathcal{A}P + \mathcal{A}PQP + \mathcal{B}\mathcal{B}^* QP + P\mathcal{A}^* + PQP\mathcal{A}^* + PQ\mathcal{B}\mathcal{B}^* - P\mathcal{C}^*\mathcal{C}P + \mathcal{B}\mathcal{B}^*$$

$$= \mathcal{A}^* P + P\mathcal{A}^* + \mathcal{B}\mathcal{B}^* + (\mathcal{A}P + \mathcal{B}\mathcal{B}^*)QP - P\mathcal{C}^*\mathcal{C}P + PQ(P\mathcal{A}^* + \mathcal{B}\mathcal{B}^*)$$

$$= 0 - P\mathcal{A}^* QP - P\mathcal{C}^*\mathcal{C}P + PQ(P\mathcal{A}^* + \mathcal{B}\mathcal{B}^*)$$

$$= -P(\mathcal{A}^* Q + \mathcal{C}^*\mathcal{C})P + PQ(P\mathcal{A}^* + \mathcal{B}\mathcal{B}^*)$$

$$= P(Q\mathcal{A})P + PQ(P\mathcal{A}^* + \mathcal{B}\mathcal{B}^*)$$

$$= PQ(\mathcal{A}P + P\mathcal{A}^* + \mathcal{B}\mathcal{B}^*)$$

$$= 0,$$

which shows the second identity. Since

$$A_L - B(I + D^* D)^{-1} B^* Y$$

$$= \mathcal{A} + \mathcal{B}\mathcal{B}^* Q(I + PQ)^{-1}$$

$$-\mathcal{B}(I + \mathcal{D}^*\mathcal{D})^{1/2}(I + \mathcal{D}^*\mathcal{D})^{-1}(I + \mathcal{D}^*\mathcal{D})^{1/2}\mathcal{B}^*(I + QP)^{-1}Q$$

$$= \mathcal{A},$$

which is stable and

$$A_L - ZC^*(I + DD^*)^{-1}C$$

$$= \mathcal{A} + \mathcal{B}\mathcal{B}^* Q(I + PQ)^{-1}$$

$$-P(I+QP)(I+PQ)^{-*}\mathcal{C}^*(I+\mathcal{D}\mathcal{D}^*)^{1/2}(I+\mathcal{D}\mathcal{D}^*)^{-1}(I+\mathcal{D}\mathcal{D}^*)^{1/2}\mathcal{C}(I+PQ)^{-1}$$

$$= \mathcal{A} + \mathcal{B}\mathcal{B}^* Q(I + PQ)^{-1} - P\mathcal{C}^*\mathcal{C}(I + PQ)^{-1}$$

$$= (I + PQ)[\mathcal{A} + (I + PQ)^{-1}P\mathcal{C}^*\mathcal{C}](I + PQ)^{-1} - P\mathcal{C}^*\mathcal{C}(I + PQ)^{-1}$$

$$= (I + PQ)\mathcal{A}(I + PQ)^{-1},$$

is stable, where we have used Lemma 2.5, we have shown that Z, Y are the stabilizing solutions to the Riccati equations. □

We are now in a position to state and prove the main theorem of this section. This theorem shows that the L-characteristic map is bijective.

THEOREM 2.10. *The map*

$$\chi_L : L_n^{p,m} \to C_n^{p,m}$$

is a bijection that preserves system equivalence. We have $\chi_L^{-1} = I\chi_L$.

Proof. In Proposition 2.8 it was shown that χ_L is injective and preserves system equivalence. Therefore it remains to show that χ_L is surjective, or more precisely that $\chi_L \cdot I\chi_L$ is the identity map. Let $(\mathcal{A}, \mathcal{B}, \mathcal{C}, \mathcal{D}) \in C_n^{p,m}$ and let

$$\left(\begin{array}{c|c} A & B \\ \hline C & D \end{array}\right) = I\chi_L\left(\left(\begin{array}{c|c} \mathcal{A} & \mathcal{B} \\ \hline \mathcal{C} & \mathcal{D} \end{array}\right)\right)$$

$$= \left(\begin{array}{c|c} \mathcal{A} + \mathcal{B}(\mathcal{B}^* \mathcal{Q} + \mathcal{D}^* \mathcal{C})(I + PQ)^{-1} & \mathcal{B}(I + \mathcal{D}^* \mathcal{D})^{1/2} \\ \hline (I + \mathcal{D}\mathcal{D}^*)^{1/2} \mathcal{C}(I + PQ)^{-1} & \mathcal{D} \end{array} \right).$$

Now consider

$$\left(\begin{array}{c|c} \mathcal{A}_1 & \mathcal{B}_1 \\ \hline \mathcal{C}_1 & \mathcal{D}_1 \end{array} \right) = \chi_L \cdot I\chi_L \left(\left(\begin{array}{c|c} \mathcal{A} & \mathcal{B} \\ \hline \mathcal{C} & \mathcal{D} \end{array} \right) \right) = \chi_L \left(\left(\begin{array}{c|c} A & B \\ \hline C & D \end{array} \right) \right)$$

$$= \left(\begin{array}{c|c} A_L - B(I + D^* D)^{-1} B^* Y & B(I + D^* D)^{-1/2} \\ \hline (I + DD^*)^{-1/2} C(I + ZY) & D \end{array} \right),$$

where Z, Y are the stabilizing solutions to the Riccati equations,

$$0 = A_L^* Y + Y A_L - Y B(I + D^* D)^{-1} B^* Y + C^*(I + DD^*)^{-1} C,$$

$$0 = A_L Z + Z A_L^* - Z C^*(I + DD^*)^{-1} C Z + B(I + D^* D)^{-1} B^*.$$

By Lemma 2.9

$$Y = Q(I + PQ)^{-1} = (I + QP)^{-1} Q,$$

$$Z = P(I + QP),$$

with Q, P the positive definite solutions to

$$\mathcal{A}P + P\mathcal{A}^* = -\mathcal{B}\mathcal{B}^*, \quad \mathcal{A}^* Q + Q\mathcal{A} = -\mathcal{C}^* \mathcal{C}.$$

Now, using that $A_L = \mathcal{A} + \mathcal{B}\mathcal{B}^* Q(I + PQ)^{-1}$ (proof of Lemma 2.9), we have $\mathcal{D}_1 = \mathcal{D}$, and

$$\mathcal{A}_1 = A_L - B(I + D^* D)^{-1} B^* Y$$

$$= \mathcal{A} + \mathcal{B}\mathcal{B}^* Q(I + PQ)^{-1}$$

$$-\mathcal{B}(I + \mathcal{D}^* \mathcal{D})^{1/2} (I + \mathcal{D}^* \mathcal{D})^{-1} (I + \mathcal{D}^* \mathcal{D})^{1/2} \mathcal{B} Q(I + PQ)^{-1}$$

$$= \mathcal{A},$$

$$\mathcal{B}_1 = B(I + D^* D)^{-1/2} = \mathcal{B}(I + \mathcal{D}^* \mathcal{D})^{1/2} (I + \mathcal{D}^* \mathcal{D})^{-1/2}$$

$$= \mathcal{B},$$

$$\mathcal{C}_1 = (I + DD^*)^{-1/2} C(I + ZY)$$

$$= (I + \mathcal{D}\mathcal{D}^*)^{-1/2}(I + \mathcal{D}\mathcal{D}^*)^{1/2}\mathcal{C}(I + PQ)^{-1}(I + PQ)$$

$$= \mathcal{C},$$

which shows the claim that $\chi_L \cdot I\chi_L$ is the identity map. Therefore χ_L is invertible with $\chi_L^{-1} = I\chi_L$. □

The notion of balancing that is appropriate for minimal systems is that of LQG balancing (Jonckheere and Silverman [16]) (see Ober and McFarlane [22] for the non-strictly proper case).

DEFINITION 2.11. *A system $(A, B, C, D) \in L_n^{p,m}$ is called LQG-balanced if*

$$A_L^* \Sigma + \Sigma A_L - \Sigma B(I + D^* D)^{-1} B^* \Sigma + C^*(I + DD^*)^{-1} C = 0,$$

$$A_L \Sigma + \Sigma A_L^* - \Sigma C^*(I + DD^*)^{-1} C\Sigma + B(I + D^* D)^{-1} B^* = 0,$$

for

$$\Sigma = diag(\sigma_1 I_{n_1}, \sigma_2 I_{n_2}, \ldots, \sigma_k I_{n_k}), \quad \sigma_1 > \sigma_2 > \cdots > \sigma_k > 0.$$

The matrix Σ is called the LQG grammian *of the system.*

In Ober and McFarlane [22] (see also [21]) the following canonical form for SISO systems was given in terms of LQG balanced realizations.

THEOREM 2.12. *The following two statements are equivalent:*

(i) $g(s)$ is the transfer function of a minimal system over $\Re$ of McMillan degree n.

(ii) $g(s)$ has a standard n-dimensional realization (A, b, c, d) given by a standard set of parameters such that

$$a_{ij} = \frac{-b_i b_j}{1 + d^2}\left(\frac{1 - s_i s_j \sigma_i \sigma_j}{s_i s_j \sigma_i + \sigma_j} - s_j d\right).$$

Moreover, the realization given in (ii) is LQG balanced with LQG grammian $\Sigma = diag(\sigma_1 I_{n_1}, \sigma_2 I_{n_2}, \ldots, \sigma_k I_{n_k})$. The map which assigns to each minimal system the realization in (ii) is a canonical form.

For an analysis of the characteristic map from the point of view of balancing it is more appropriate to slightly change the definition of the characteristic map. This is done by performing a state-space transformation on $\chi_L((A, B, C, D))$.

Define for a system $(A, B, C, D) \in L_n^{p,m}$ the modified characteristic map

$$\tilde{\chi}_L\left(\left(\begin{array}{c|c} A & B \\ \hline C & D \end{array}\right)\right)$$

$$= \left(\begin{array}{c|c} T^{1/4}(A_L - B(I+D^*D)^{-1}B^*Y)T^{-1/4} & T^{1/4}B(I+D^*D)^{-1/2} \\ \hline (I+DD^*)^{-1/2}C(I+ZY)T^{-1/4} & D \end{array} \right),$$

where $T := (I+ZY)^*(I+ZY)$ and Z, Y are the stabilizing solutions to the two Riccati equations.

COROLLARY 2.13. *The map $\tilde{\chi}_L : L_n^{p,m} \to C_n^{p,m}$ is a bijection with the following properties,*

1. *$(A,B,C,D) \in L_n^{p,m}$ is LQG balanced with LQG grammian Σ if and only if $\tilde{\chi}_L((A,B,C,D))$ is Lyapunov balanced with Lyapunov grammian Σ.*
2. *$(A,b,c,d) \in L_n^{1,1}$ is in LQG-balanced canonical form of Theorem 2.12 if and only if $\tilde{\chi}_L((A,b,c,d))$ is in Lyapunov balanced canonical form of Theorem 1.1.*

Proof. 1.) Let $(A,B,C,D) \in L_n^{p,m}$ and let Y, Z be the stabilizing solutions to the two Riccati equations. By Lemma 2.7 $(\mathcal{A},\mathcal{B},\mathcal{C},\mathcal{D})=$ $\chi_L((A,B,C,D))$ is such that $P=(I+ZY)^{-1}Z$, $Q=Y(I+ZY)$ solve the Lyapunov equations,

$$\mathcal{A}P + P\mathcal{A}^* = -\mathcal{B}\mathcal{B}^*, \quad \mathcal{A}^*Q + Q\mathcal{A} = -\mathcal{C}^*\mathcal{C}.$$

If $(\mathcal{A}_1,\mathcal{B}_1,\mathcal{C}_1,\mathcal{D}_1) = \tilde{\chi}_L((A,B,C,D))$, then the positive definite solution to the Lyapunov equations,

$$\mathcal{A}_1P_1 + P_1\mathcal{A}_1^* = -\mathcal{B}_1\mathcal{B}_1^*, \quad \mathcal{A}_1^*Q_1 + Q_1\mathcal{A}_1 = -\mathcal{C}_1^*\mathcal{C}_1,$$

are given by

$$P_1 = T^{1/4}PT^{1/4}$$

$$= ((I+ZY)^*(I+ZY))^{1/4}(I+ZY)^{-1}Z\,((I+ZY)^*(I+ZY))^{1/4},$$

$$Q_1 = T^{-1/4}QT^{-1/4}$$

$$= ((I+ZY)^*(I+ZY))^{-1/4}Y(I+ZY)\,((I+ZY)^*(I+ZY))^{-1/4}.$$

If (A,B,C,D) is LQG balanced then $Z=Y=\Sigma$ is diagonal and therefore

$$P_1 = Z = Y = Q_1,$$

i.e. $(\mathcal{A}_1,\mathcal{B}_1,\mathcal{C}_1,\mathcal{D}_1)$ is Lyapunov balanced with Lyapunov grammian Σ. The converse follows in the same way.

2.) This follows by straightforward verification. □

3. Bounded real systems. The next class of systems that we will consider are bounded real systems. We call a system *bounded real*, if it is stable and its transfer function satisfies,

$$I - G^*(i\omega)G(i\omega) > 0,$$

for all $\omega \in \Re \cup \{\pm\infty\}$. We denote by $B_n^{p,m} \subseteq L_n^{p,m}$ the subset of bounded real systems. We will proceed as in the previous section and construct a map of the set of bounded real systems into the set of stable systems. Whereas in the previous case the map was onto, this is not the case here. In the present case the map will be a bijection between the class of bounded real systems of McMillan degree n and the set of stable minimal system of the same McMillan degree whose Hankel singular values are less than one.

To simplify presentation, we are going to use the following notation,

$$A_B := A - B(I - D^*D)^{-1}D^*C.$$

We are now going to define the B-characteristic of a bounded real system.

DEFINITION 3.1. *Let* $\left(\begin{array}{c|c} A & B \\ \hline C & D \end{array}\right)$ *be a minimal bounded real system. Let* Y *be the stabilizing solution of the Riccati equation*

$$0 = A_B^*Y + YA_B + YB(I - D^*D)^{-1}B^*Y + C^*(I - DD^*)^{-1}C,$$

i.e. $A_B + B(I - D^*D)^{-1}B^*Y$ *is stable, and let* Z *be the stabilizing solution to the Riccati equation*

$$0 = A_BZ + ZA_B^* + ZC^*(I - DD^*)^{-1}CZ + B(I - D^*D)^{-1}B^*,$$

i.e. $A_B + ZC^*(I - DD^*)^{-1}C$ *is stable.*

Then the system

$$\chi_B\left(\left(\begin{array}{c|c} A & B \\ \hline C & D \end{array}\right)\right) := \left(\begin{array}{c|c} A_B + B(I - D^*D)^{-1}B^*Y & B(I - D^*D)^{-1/2} \\ \hline -(I - DD^*)^{-1/2}C(I - ZY) & D \end{array}\right)$$

is called the B-characteristic *of the system* $\left(\begin{array}{c|c} A & B \\ \hline C & D \end{array}\right)$.

These following relations are the relations that are equivalent to the Bucy relations for the case of minimal systems. Note that standard results on the bounded real Riccati equation ([24]) imply that $I - ZY$ is nonsingular, where Y and Z are the stabilizing solutions to the two bounded real Riccati equations.

LEMMA 3.2. *Let* $\left(\begin{array}{c|c} A & B \\ \hline C & D \end{array}\right)$ *be a minimal bounded real system. Let* Y *be a solution of the Riccati equation*

$$0 = A_B^*Y + YA_B + YB(I - D^*D)^{-1}B^*Y + C^*(I - DD^*)^{-1}C,$$

and let Z be a solution of the Riccati equation

$$0 = A_B Z + Z A_B^* + Z C^* (I - D D^*)^{-1} C Z + B(I - D^* D)^{-1} B^*,$$

then

$$[I - ZY](A_B + B(I - D^* D)^{-1} B^* Y) = (A_B + Z C^* (I - D D^*)^{-1} C)[I - ZY].$$

Proof. Consider the two Riccati equations,

$$0 = A_B^* Y + Y A_B + Y B(I - D^* D)^{-1} B^* Y + C^* (I - D D^*)^{-1} C,$$

and

$$0 = A_B Z + Z A_B^* + Z C^* (I - D D^*)^{-1} C Z + B(I - D^* D)^{-1} B^*.$$

Multiplying the first equation on the left by $-Z$ and the second equation on the right by $-Y$, equating both equations and adding A_B to both sides we obtain

$$A_B - Z A_B^* Y - Z Y A_B - Z Y B(I - D^* D)^{-1} B^* Y - Z C^* (I - D D^*)^{-1} C$$

$$= A_B - A_B Z Y - Z A_B^* Y - Z C^* (I - D D^*)^{-1} C Z Y - B(I - D^* D)^{-1} B^* Y.$$

Canceling the term $Z A_B^* Y$ from either side and collecting terms, we obtain

$$[I - ZY](A_B + B(I - D^* D)^{-1} B^* Y) = (A_B + Z C^* (I - D D^*)^{-1} C)[I - ZY].$$

□

As a consequence of these Bucy type relationships we can rewrite the B-characteristic of a bounded-real system as follows,

$$\chi_B\left(\left(\begin{array}{c|c} A & B \\ \hline C & D \end{array}\right)\right) :=$$

$$\left(\begin{array}{c|c} [I - ZY]^{-1}(A_B + Z C^* (I - D D^*)^{-1} C)[I - ZY]) & B(I - D^* D)^{-1/2} \\ \hline -(I - D D^*)^{-1/2} C(I - ZY) & D \end{array}\right)$$

The following Lemma shows that the B-characteristic map maps a bounded real system to a stable system of the same McMillan degree.

LEMMA 3.3. *The B-characteristic of a minimal bounded-real system is stable and minimal. The B-characteristics of two equivalent systems are equivalent.*

Proof. Since Y is the stabilizing solution of the Riccati equation the matrix $A_B + B(I - D^* D)^{-1} B^* Y$ is stable by definition. It is easily seen that the characteristic system is reachable. The observability of the system follows by using the representation of the characteristic in which the A

matrix has been written in the form resulting from the Bucy type relations. That the B-characteristics of two equivalent systems are equivalent is easily verified. □

In the following Lemma we investigate the solutions of the Lyapunov equations of the B-characteristic system.

LEMMA 3.4. *Let* $\left(\begin{array}{c|c} A & B \\ \hline C & D \end{array}\right)$ *be a minimal bounded-real system and let*

$$\left(\begin{array}{c|c} \mathcal{A} & \mathcal{B} \\ \hline \mathcal{C} & \mathcal{D} \end{array}\right) := \left(\begin{array}{c|c} A_B + B(I-D^*D)^{-1}B^*Y & B(I-D^*D)^{-1/2} \\ \hline -(I-DD^*)^{-1/2}C(I-ZY) & D \end{array}\right)$$

be its B-characteristic system, with Y and Z the stabilizing solutions to the respective Riccati equations. Then the Lyapunov equations

$$\mathcal{A}P + P\mathcal{A}^* = -\mathcal{B}\mathcal{B}^*$$

$$\mathcal{A}^*Q + Q\mathcal{A} = -\mathcal{C}^*\mathcal{C}$$

have solutions given by

$$P = (I-ZY)^{-1}Z = Z(I-YZ)^{-1}$$

$$Q = Y(I-ZY) = (I-YZ)Y.$$

Proof. We first show that $\mathcal{A}P + P\mathcal{A}^* = -\mathcal{B}\mathcal{B}^*$. Since $I-ZY$ is invertible this follows from,

$$(I-ZY)(\mathcal{A}P + P\mathcal{A}^*)(I-YZ)$$

$$= (I-ZY)\mathcal{A}Z + Z\mathcal{A}^*(I-YZ)$$

$$= (I-ZY)[A_B+B(I-D^*D)^{-1}B^*Y]Z+Z[A_B+B(I-D^*D)^{-1}B^*Y]^*(I-YZ)$$

$$= A_BZ + ZA_B^* - Z(YA_B + A_B^*Y)Z$$

$$+(I-ZY)B(I-D^*D)^{-1}B^*YZ + ZYB(I-D^*D)^{-1}B^*(I-YZ)$$

$$= -ZC^*(I-DD^*)^{-1}CZ - B(I-D^*D)^{-1}B^*$$

$$-Z[-YB(I-D^*D)^{-1}B^*Y - C^*(I-DD^*)^{-1}C]Z$$

$$+(I-ZY)B(I-D^*D)^{-1}B^*YZ+ZYB(I-D^*D)^{-1}B^*(I-YZ)$$

$$=-B(I-D^*D)^{-1}B^*+ZYB(I-D^*D)^{-1}B^*YZ$$

$$+(I-ZY)B(I-D^*D)^{-1}B^*YZ+ZYB(I-D^*D)^{-1}B^*(I-YZ)$$

$$=-(I-ZY)B(I-D^*D)^{-1}B^*(I-YZ)$$

$$=-(I-ZY)\mathcal{B}\mathcal{B}^*(I-YZ).$$

Let now $Q=Y-YZY$. We are going to show that

$$\mathcal{A}^*Q+Q\mathcal{A}=-\mathcal{C}^*\mathcal{C}.$$

We consider

$$\mathcal{A}^*Q+Q\mathcal{A}$$

$$=[A_B+B(I-D^*D)^{-1}B^*Y]^*Y(I-ZY)+(I-YZ)Y[A_B+B(I-D^*D)^{-1}B^*Y]$$

$$=[A_B^*Y+YB(I-D^*D)^{-1}B^*Y](I-ZY)$$

$$+(I-YZ)[YA_B+YB(I-D^*D)^{-1}B^*Y],$$

using the bounded real Riccati equation this gives,

$$=[-YA_B-C^*(I-DD^*)^{-1}C](I-ZY)+(I-YZ)[-A_B^*Y-C^*(I-DD^*)^{-1}C]$$

$$=-C^*(I-DD^*)^{-1}C(I-ZY)-(I-YZ)C^*(I-DD^*)^{-1}C$$

$$-YA_B-A_B^*Y+Y(A_BZ+ZA_B^*)Y$$

$$=-C^*(I-DD^*)^{-1}(I-ZY)-(I-YZ)C^*(I-DD^*)^{-1}C$$

$$+YB(I-D^*D)^{-1}B^*Y+C^*(I-DD^*)^{-1}C$$

$$+Y[-ZC^*(I-DD^*)^{-1}CZ-B(I-D^*D)^{-1}B^*]Y$$

$$=-(I-YZ)C^*(I-DD^*)^{-1}C(I-ZY)$$

$$=-\mathcal{C}^*\mathcal{C}.$$

□

A consequence of the previous Lemma is that the image of χ_B is not the whole set $C_n^{p,m}$, but the subset $UC_{n,B}^{p,m}$. A system $(\mathcal{A},\mathcal{B},\mathcal{C},\mathcal{D})\in C_n^{p,m}$ is in $UC_{n,B}^{p,m}$ if

1. $\lambda_{max}(PQ) < 1$, where P, Q are the positive definite solutions to

$$\mathcal{A}P + P\mathcal{A}^* = -\mathcal{B}\mathcal{B}^*, \quad \mathcal{A}^*Q + Q\mathcal{A} = -\mathcal{C}^*\mathcal{C},$$

2. $I - \mathcal{D}^*\mathcal{D} > 0$.

We can now show that the B-characteristic of a bounded real system is minimal and in $UC^{p,m}_{n,B}$.

LEMMA 3.5. *We have*

$$\chi_B(B^{p,m}_n) \subseteq UC^{p,m}_{n,B}.$$

Proof. Let $(A,B,C,D) \in B^{p,m}_n$. It follows from standard results on the bounded real Riccati equations that $\lambda_{max}(ZY) < 1$ ([24]), where Z, Y are the stabilizing solutions to the bounded real Riccati equations. If $(\mathcal{A},\mathcal{B},\mathcal{C},\mathcal{D}) = \chi_B((A,B,C,D))$ and P, Q are the positive definite solutions to the Lyapunov equations

$$\mathcal{A}P + P\mathcal{A}^* = -\mathcal{B}\mathcal{B}^*, \quad \mathcal{A}^*Q + Q\mathcal{A} = -\mathcal{C}^*\mathcal{C},$$

then as a consequence of Lemma 3.4 we have $PQ = ZY$ and therefore that $\lambda_{max}(PQ) < 1$. Clearly, $I - \mathcal{D}^*\mathcal{D} = I - D^*D > 0$ and hence $(\mathcal{A},\mathcal{B},\mathcal{C},\mathcal{D}) \in UC^{p,m}_{n,B}$. □

In the next definition we are going to define the inverse B-characteristic map $I\chi_B : UC^{p,m}_{n,B} \to B^{p,m}_n$. Analogously to the case of minimal systems $I\chi_B$ will turn out to be the inverse of the B-characteristic map $\chi_B : B^{p,m}_n \to UC^{p,m}_{n,B}$.

DEFINITION 3.6. *Let* $(\mathcal{A},\mathcal{B},\mathcal{C},\mathcal{D}) \in UC^{p,m}_{n,B}$ *and let* P, Q *be the solutions to the Lyapunov equations*

$$\mathcal{A}P + P\mathcal{A}^* = -\mathcal{B}\mathcal{B}^*, \quad \mathcal{A}^*Q + Q\mathcal{A} = -\mathcal{C}^*\mathcal{C}.$$

Then

$$I\chi_B\left(\left(\begin{array}{c|c} \mathcal{A} & \mathcal{B} \\ \hline \mathcal{C} & \mathcal{D} \end{array}\right)\right)$$

$$:= \left(\begin{array}{c|c} \mathcal{A} - \mathcal{B}(\mathcal{B}^*Q + \mathcal{D}^*\mathcal{C})(I - PQ)^{-1} & \mathcal{B}(I - \mathcal{D}^*\mathcal{D})^{1/2} \\ \hline -(I - \mathcal{D}\mathcal{D}^*)^{1/2}\mathcal{C}(I - PQ)^{-1} & \mathcal{D} \end{array}\right)$$

is called the inverse B-characteristic system.

In order to be able to show that $I\chi_B$ maps a stable system in $UC^{p,m}_{n,B}$ to a bounded real system we need the following Lemmas.

LEMMA 3.7. *Let* $(\mathcal{A},\mathcal{B},\mathcal{C},\mathcal{D}) \in UC^{p,m}_{n,B}$ *and let* P *and* Q *be such that*

$$\mathcal{A}P + P\mathcal{A}^* = -\mathcal{B}\mathcal{B}^*, \quad \mathcal{A}^*Q + Q\mathcal{A} = -\mathcal{C}^*\mathcal{C}.$$

Then

$$[\mathcal{A} - \mathcal{B}(\mathcal{B}^*Q + \mathcal{D}^*\mathcal{C})(I - PQ)^{-1}][I - PQ]$$

$$= [I - PQ][\mathcal{A} - (I - PQ)^{-1}(P\mathcal{C}^* + \mathcal{B}\mathcal{D}^*)\mathcal{C}].$$

Proof. We have

$$[\mathcal{A} - \mathcal{B}(\mathcal{B}^* Q + \mathcal{D}^*\mathcal{C})(I - PQ)^{-1}][I - PQ]$$

$$= \mathcal{A}(I - PQ) - \mathcal{B}(\mathcal{B}^* Q + \mathcal{D}^*\mathcal{C}) = \mathcal{A} - \mathcal{A}PQ - \mathcal{B}\mathcal{B}^* Q - \mathcal{B}\mathcal{D}^*\mathcal{C}$$

$$= \mathcal{A} - [\mathcal{A}P + \mathcal{B}\mathcal{B}^*]Q - \mathcal{B}\mathcal{D}^*\mathcal{C}$$

$$= \mathcal{A} - [-P\mathcal{A}^*]Q - \mathcal{B}\mathcal{D}^*\mathcal{C} = \mathcal{A} - PQ\mathcal{A} - P\mathcal{C}^*\mathcal{C} - \mathcal{B}\mathcal{D}^*\mathcal{C}$$

$$= [I - PQ][\mathcal{A} - (I - PQ)^{-1}(P\mathcal{C}^* + \mathcal{B}\mathcal{D}^*)\mathcal{C}].$$

□

LEMMA 3.8. *Let $(\mathcal{A}, \mathcal{B}, \mathcal{C}, \mathcal{D}) \in UC^{p,m}_{n,B}$. Let P, Q be the positive definite solutions to the Lyapunov equations*

$$\mathcal{A}P + P\mathcal{A}^* + \mathcal{B}\mathcal{B}^* = 0, \quad \mathcal{A}^*Q + Q\mathcal{A} + \mathcal{C}^*\mathcal{C} = 0.$$

Let

$$\left(\begin{array}{c|c} A & B \\ \hline C & D \end{array}\right) := I\chi_B\left(\left(\begin{array}{c|c} \mathcal{A} & \mathcal{B} \\ \hline \mathcal{C} & \mathcal{D} \end{array}\right)\right).$$

Then

$$0 = A_B^* Y + YA_B + YB(I - D^*D)^{-1}B^*Y + C^*(I - DD^*)^{-1}C,$$

$$0 = A_B Z + ZA_B^* + ZC^*(I - DD^*)^{-1}CZ + B(I - D^*D)^{-1}B^*,$$

with

$$Y = Q(I - PQ)^{-1} = (I - QP)^{-1}Q,$$

$$Z = (I - PQ)P = P(I - QP).$$

Moreover, Y and Z are the stabilizing solutions to these two bounded real Riccati equations.

Proof. First note that

$$A_B = A - B(I - D^*D)^{-1}D^*C$$

$$= \mathcal{A} - \mathcal{B}(\mathcal{B}^*Q + \mathcal{D}^*\mathcal{C})(I - PQ)^{-1}$$

$$+\mathcal{B}(I-\mathcal{D}^*\mathcal{D})^{1/2}(I-\mathcal{D}^*\mathcal{D})^{-1}\mathcal{D}^*(I-\mathcal{D}\mathcal{D}^*)^{1/2}\mathcal{C}(I-PQ)^{-1}$$

$$=\mathcal{A}-\mathcal{B}\mathcal{B}^*Q(I-PQ)^{-1}.$$

Since $I-QP$ is invertible and

$$(I-QP)[A_B^*Y+YA_B+YB(I-D^*D)^{-1}B^*Y+C^*(I-DD^*)^{-1}C](I-PQ)$$

$$=(I-QP)[(\mathcal{A}-\mathcal{B}\mathcal{B}^*Q(I-PQ)^{-1})^*Q(I-PQ)^{-1}$$

$$+(I-QP)^{-1}Q(\mathcal{A}-\mathcal{B}\mathcal{B}^*Q(I-PQ)^{-1})$$

$$+(I-QP)^{-1}Q\mathcal{B}(I-\mathcal{D}^*\mathcal{D})^{1/2}(I-\mathcal{D}^*\mathcal{D})^{-1}(I-\mathcal{D}^*\mathcal{D})^{1/2}\mathcal{B}^*(I-PQ)^{-1}$$

$$+(I-PQ)^{-*}\mathcal{C}^*(I-\mathcal{D}\mathcal{D}^*)^{1/2}(I-\mathcal{D}\mathcal{D}^*)^{-1}(I-\mathcal{D}\mathcal{D}^*)^{1/2}\mathcal{C}(I-PQ)^{-1}](I-PQ)$$

$$=(I-QP)\mathcal{A}^*Q-Q\mathcal{B}\mathcal{B}^*Q+Q\mathcal{A}(I-PQ)-Q\mathcal{B}\mathcal{B}^*Q+Q\mathcal{B}\mathcal{B}^*Q+\mathcal{C}^*\mathcal{C}$$

$$=\mathcal{A}^*Q+Q\mathcal{A}+\mathcal{C}^*\mathcal{C}-Q(P\mathcal{A}^*+\mathcal{A}P+\mathcal{B}\mathcal{B}^*)Q$$

$$=0,$$

we have verified the first identity. Now with $Z=P(I-QP)$ we have

$$A_BZ+ZA_B^*+ZC^*(I-DD^*)^{-1}CZ+B(I-D^*D)^{-1}B^*$$

$$=(\mathcal{A}-\mathcal{B}\mathcal{B}^*Q(I-PQ)^{-1})(I-PQ)P+P(I-PQ)(\mathcal{A}-\mathcal{B}\mathcal{B}^*Q(I-PQ)^{-1})^*$$

$$+P(I-QP)(I-PQ)^{-*}\mathcal{C}^*\mathcal{C}(I-PQ)^{-1}(I-PQ)P$$

$$+\mathcal{B}(I-\mathcal{D}^*\mathcal{D})^{1/2}(I-\mathcal{D}^*\mathcal{D})^{-1}(I-\mathcal{D}^*\mathcal{D})^{1/2}\mathcal{B}^*$$

$$=\mathcal{A}P-\mathcal{A}PQP-\mathcal{B}\mathcal{B}^*QP+P\mathcal{A}^*-PQP\mathcal{A}^*-PQ\mathcal{B}\mathcal{B}^*+P\mathcal{C}^*\mathcal{C}P+\mathcal{B}\mathcal{B}^*$$

$$=\mathcal{A}^*P+P\mathcal{A}^*+\mathcal{B}\mathcal{B}^*-(\mathcal{A}P+\mathcal{B}\mathcal{B}^*)QP+P\mathcal{C}^*\mathcal{C}P-PQ(P\mathcal{A}^*+\mathcal{B}\mathcal{B}^*)$$

$$=0+P\mathcal{A}^*QP+P\mathcal{C}^*\mathcal{C}P-PQ(P\mathcal{A}^*+\mathcal{B}\mathcal{B}^*)$$

$$= P(\mathcal{A}^*Q + \mathcal{C}^*\mathcal{C})P - PQ(P\mathcal{A}^* + \mathcal{B}\mathcal{B}^*)$$

$$= -P(Q\mathcal{A})P - PQ(P\mathcal{A}^* + \mathcal{B}\mathcal{B}^*)$$

$$= -PQ(\mathcal{A}P + P\mathcal{A}^* + \mathcal{B}\mathcal{B}^*)$$

$$= 0,$$

which shows the second identity. Since

$$A_B + B(I - D^*D)^{-1}B^*Y$$

$$= \mathcal{A} - \mathcal{B}\mathcal{B}^*Q(I - PQ)^{-1}$$

$$+\mathcal{B}(I - \mathcal{D}^*\mathcal{D})^{1/2}(I - \mathcal{D}^*\mathcal{D})^{-1}(I - \mathcal{D}^*\mathcal{D})^{1/2}\mathcal{B}^*Q(I - PQ)^{-1}$$

$$= \mathcal{A},$$

is stable and

$$A_B + ZC^*(I - DD^*)^{-1}C$$

$$= \mathcal{A} - \mathcal{B}\mathcal{B}^*Q(I - PQ)^{-1}$$

$$+P(I-QP)(I-PQ)^{-*}\mathcal{C}^*(I-\mathcal{D}\mathcal{D}^*)^{1/2}(I-\mathcal{D}\mathcal{D}^*)^{-1}(I-\mathcal{D}\mathcal{D}^*)^{1/2}\mathcal{C}(I-PQ)^{-1}$$

$$= \mathcal{A} - \mathcal{B}\mathcal{B}^*Q(I - PQ)^{-1} + P\mathcal{C}^*\mathcal{C}(I - PQ)^{-1}$$

$$= (I - PQ)[\mathcal{A} - (I - PQ)^{-1}P\mathcal{C}^*\mathcal{C}](I + PQ)^{-1} + P\mathcal{C}^*\mathcal{C}(I + PQ)^{-1}$$

$$= (I - PQ)\mathcal{A}(I - PQ)^{-1},$$

is stable, where we have used Lemma 3.7, we have shown that Z, Y are the stabilizing solutions to the bounded real Riccati equations. □

We can now show that $I\chi_B$ maps $UC^{p,m}_{n,B}$ into $B^{p,m}_n$.

LEMMA 3.9. *We have*

$$I\chi_B(UC^{p,m}_{n,B}) \subseteq B^{p,m}_n.$$

Proof. Let $(\mathcal{A},\mathcal{B},\mathcal{C},\mathcal{D}) \in UC_{n,B}^{p,m}$ and let $(A,B,C,D) = I\chi_B((\mathcal{A},\mathcal{B},\mathcal{C},\mathcal{D}))$. It follows immediately that (A,B,C,D) is reachable. After rewriting of (A,B,C,D) using the formula of Lemma 3.7 it follows that the system is observable. In Lemma 3.8 it was shown that the two bounded real Riccati equations for the system (A,B,C,D) have positive definite stabilizing solutions Y, Z. This together with the fact that $\lambda_{max}(YZ) = \lambda_{max}(PQ) < 1$ implies ([24]) that (A,B,C,D) is bounded real. □

We are now in a position to show that $\chi_B : B_n^{p,m} \to UC_{n,B}^{p,m}$ is a bijection.

THEOREM 3.10. *The map*

$$\chi_B : B_n^{p,m} \to UC_{n,B}^{p,m}$$

is a bijection that preserves system equivalence. The inverse map is given by $\chi_B^{-1} = I\chi_B$.

Proof. That χ_B preserves system equivalence was established in Lemma 3.3.We are next going to show that χ_B is injective, or more precisely that $I\chi_B \cdot \chi_B$ is the identity map. Let $(A,B,C,D) \in B_n^{p,m}$, let $(\mathcal{A},\mathcal{B},\mathcal{C},\mathcal{D}) = \chi_B((A,B,C,D))$ and set $(A_1,B_1,C_1,D_1) := I\chi_B((\mathcal{A},\mathcal{B},\mathcal{C},\mathcal{D}))$. Using Lemma 3.4 we have

$$D_1 = D,$$

$$B_1 = \mathcal{B}(I-\mathcal{D}^*\mathcal{D})^{1/2} = B(I-D^*D)^{-1/2}(I-D^*D)^{1/2} = B,$$

$$C_1 = -(I-\mathcal{D}\mathcal{D}^*)^{1/2}\mathcal{C}(I-PQ)^{-1}$$

$$= (I-DD^*)^{-1/2}(I-DD^*)^{1/2}C(I-ZY)(I-ZY)^{-1} = C,$$

$$A_1 = \mathcal{A} - \mathcal{B}(\mathcal{B}^*Q + \mathcal{D}^*\mathcal{C})(I-PQ)^{-1}$$

$$= A_B + B(I-D^*D)^{-1}B^*Y - B(I-D^*D)^{-1/2}(I-D^*D)^{-1/2}B^*Y$$

$$+B(I-D^*D)^{-1/2}D^*(I-DD^*)^{-1/2}C(I-ZY)(I-ZY)^{-1}$$

$$= A - B(I-D^*D)^{-1}D^*C + B(I-D^*D)^{-1}D^*C$$

$$= A,$$

which shows that $I\chi_B \cdot \chi_B((A,B,C,D)) = (A,B,C,D)$ and hence that χ_B is injective. We now show that χ_B is surjective by showing that $\chi_B \cdot$

$I\chi_B$ is the identity map. Let $(\mathcal{A},\mathcal{B},\mathcal{C},\mathcal{D}) \in UC^{p,m}_{n,B}$, let $(A,B,C,D) := I\chi_B((\mathcal{A},\mathcal{B},\mathcal{C},\mathcal{D}))$ and set $(\mathcal{A}_1,\mathcal{B}_1,\mathcal{C}_1,\mathcal{D}_1) := \chi_B((A,B,C,D))$. Then

$$\mathcal{D}_1 = \mathcal{D},$$

$$\mathcal{C}_1 = -(I - D^*D)^{-1/2}C(I - ZY)$$

$$= (I - \mathcal{D}\mathcal{D}^*)^{-1/2}(I - \mathcal{D}\mathcal{D}^*)^{1/2}\mathcal{C}(I - PQ)^{-1}(I - PQ)$$

$$= \mathcal{C},$$

$$\mathcal{B}_1 = B(I - D^*D)^{-1/2} = \mathcal{B}(I - \mathcal{D}^*\mathcal{D})^{1/2}(I - \mathcal{D}^*\mathcal{D})^{-1/2} = \mathcal{B},$$

$$\mathcal{A}_1 = A_B + B(I - D^*D)^{-1}B^*Y$$

$$= \mathcal{A} - \mathcal{B}\mathcal{B}^*Q(I - PQ)^{-1}$$

$$+\mathcal{B}(I - \mathcal{D}^*\mathcal{D})^{1/2}(I - \mathcal{D}^*\mathcal{D})^{-1}(I - \mathcal{D}^*\mathcal{D})^{1/2}\mathcal{B}^*Q(I - PQ)^{-1}$$

$$= \mathcal{A}.$$

This shows that χ_B is surjective. Hence we have that χ_B is bijective with inverse $\chi_B^{-1} = I\chi_B$. □

We now come to analyze the previous result from the point of view of balanced realizations. Bounded real balanced realizations were introduced by Opdenacker and Jonckheere [23].

DEFINITION 3.11. *A system* $(A,B,C,D) \in B^{p,m}_n$ *is called* bounded real balanced *if*

$$A_B^*\Sigma + \Sigma A_B + \Sigma B(I - D^*D)^{-1}B^*\Sigma + C^*(I - DD^*)^{-1}C = 0,$$

$$A_B\Sigma + \Sigma A_B^* + \Sigma C^*(I - DD^*)^{-1}C\Sigma + B(I - D^*D)^{-1}B^* = 0,$$

for

$$\Sigma = diag(\sigma_1 I_{n_1}, \sigma_2 I_{n_2}, \ldots, \sigma_k I_{n_k}), \quad \sigma_1 > \sigma_2 > \cdots > \sigma_k > 0,$$

and Σ *is the stabilizing solution to both equations. The matrix* Σ *is called the* bounded real grammian *of the system.*

In Ober [21] the following canonical form for SISO bounded real systems was given.

THEOREM 3.12. *The following two statements are equivalent:*

(i) $g(s)$ *is the transfer function of a bounded real system over* $\Re$ *of McMillan degree* n.

(ii) $g(s)$ *has a standard* n*-dimensional realization* (A, b, c, d) *given by a standard set of parameters such that*

$$a_{ij} = \frac{-b_i b_j}{1-d^2}\left(\frac{1+s_i s_j \sigma_i \sigma_j}{s_i s_j \sigma_i + \sigma_j} + s_j d\right),$$

with $|d| < 1$, $\sigma_1 < 1$.

Moreover, the realization given in (ii) is bounded real balanced with bounded real grammian $\Sigma = diag(\sigma_1 I_{n_1}, \sigma_2 I_{n_2}, \ldots, \sigma_k I_{n_k})$. *The map which assigns to each bounded real system the realization in (ii) is a canonical form.*

As in the case of minimal systems we now introduce a slightly modified characteristic map. Define for a system $(A, B, C, D) \in B_n^{p,m}$ the modified characteristic map

$$\tilde{\chi}_B\left(\left(\begin{array}{c|c} A & B \\ \hline C & D \end{array}\right)\right)$$

$$= \left(\begin{array}{c|c} T^{1/4}(A_B + B(I-D^*D)^{-1}B^*Y)T^{-1/4} & T^{1/4}B(I-D^*D)^{-1/2} \\ \hline -(I-DD^*)^{-1/2}C(I-ZY)T^{-1/4} & D \end{array}\right),$$

where $T := (I-ZY)^*(I-ZY)$ and Z, Y are the stabilizing solutions to the two bounded real Riccati equations. We have the following corollary.

COROLLARY 3.13. *The map* $\tilde{\chi}_B : B_n^{p,m} \to UC_{n,B}^{p,m}$ *is a bijection with the following properties,*

1. $(A, B, C, D) \in B_n^{p,m}$ *is bounded real balanced with bounded real grammian* Σ *if and only if* $\tilde{\chi}_B((A, B, C, D))$ *is Lyapunov balanced with Lyapunov grammian* Σ.
2. $(A, b, c, d) \in B_n^{1,1}$ *is in the bounded real balanced canonical form of Theorem 3.12 if and only if* $\tilde{\chi}_B((A, b, c, d))$ *is in Lyapunov balanced canonical form of Theorem 1.1.*

4. Positive real systems. We are now going to consider positive real systems. We call a square minimal system positive real, if it is stable and its transfer function satisfies,

$$G(i\omega) + G^*(i\omega) > 0,$$

for all $\omega \in \Re \cup \{\pm\infty\}$. We denote by P_n^m the subset of $L_n^{m,m}$ of positive real systems. We again introduce some notation to simplify the presentation. Let (A, B, C, D) be a positive real system, then set,

$$A_P := A - B(D+D^*)^{-1}C^*.$$

The P-characteristic of a positive real system is defined as follows.

DEFINITION 4.1. *Let* $\left(\begin{array}{c|c} A & B \\ \hline C & D \end{array}\right)$ *be a minimal positive real system. Let* Y *be the stabilizing solution of the Riccati equation*

$$0 = A_P^* Y + Y A_P + Y B(D+D^*)^{-1} B^* Y + C^*(D+D^*)^{-1} C,$$

i.e. $A_P + B(D+D^*)^{-1}B^*Y$ *is stable, and let* Z *be the stabilizing solution to the Riccati equation*

$$0 = A_P Z + Z A_P^* + Z C^*(D+D^*)^{-1} C Z + B(D+D^*)^{-1} B^*,$$

i.e. $A_P + ZC^*(D+D^*)^{-1}C$ *is stable.*

Then the system

$$\chi_P\left(\left(\begin{array}{c|c} A & B \\ \hline C & D \end{array}\right)\right) := \left(\begin{array}{c|c} A_P + B(D+D^*)^{-1}B^*Y & B(D+D^*)^{-1/2} \\ \hline -(D+D^*)^{-1/2}C(I-ZY) & D \end{array}\right)$$

is called the P-characteristic *of the system* $\left(\begin{array}{c|c} A & B \\ \hline C & D \end{array}\right)$.

The following relations are analogous to the Bucy relations for the case of minimal systems. Note that standard results on the positive real Riccati equation (see e.g. [24]) imply that $I - ZY$ is non-singular, where Y and Z are the stabilizing solutions to the two positive real Riccati equations.

LEMMA 4.2. *Let* $\left(\begin{array}{c|c} A & B \\ \hline C & D \end{array}\right)$ *be a minimal positive-real system. Let* Y *be a solution of the Riccati equation*

$$0 = A_P^* Y + Y A_P + Y B(D+D^*)^{-1} B^* Y + C^*(D+D^*)^{-1} C,$$

and let Z *be a solution of the Riccati equation*

$$0 = A_P Z + Z A_P^* + Z C^*(D+D^*)^{-1} C Z + B(D+D^*)^{-1} B^*,$$

then

$$[I - ZY](A_P + B(D+D^*)^{-1}B^*Y) = (A_P + ZC^*(D+D^*)^{-1}C)[I - ZY].$$

Proof. Consider the two Riccati equations,

$$0 = A_P^* Y + Y A_P + Y B(D+D^*)^{-1} B^* Y + C^*(D+D^*)^{-1} C,$$

and

$$0 = A_P Z + Z A_P^* + Z C^*(D+D^*)^{-1} C Z + B(D+D^*)^{-1} B^*.$$

Multiplying the first equation on the left by $-Z$ and the second equation on the right by $-Y$, equating both equations and adding A_P to both sides we obtain

$$A_P - Z A_P^* Y - Z Y A_P - Z Y B(D+D^*)^{-1} B^* Y - Z C^*(D+D^*)^{-1} C$$

$$= A_P - A_P ZY - ZA_P^* Y - ZC^*(D+D^*)^{-1}CZY - B(D+D^*)^{-1}B^*Y.$$

Canceling the term ZA_P^*Y from either side and collecting terms, we obtain

$$[I - ZY](A_P + B(D+D^*)^{-1}B^*Y) = (A_P + ZC^*(D+D^*)^{-1}C)[I - ZY].$$

□

As a consequence of these Bucy type relationships we can rewrite the P-characteristic of a positive real system as follows,

$$\chi_P\left(\left(\begin{array}{c|c} A & B \\ \hline C & D \end{array}\right)\right) :=$$

$$\left(\begin{array}{c|c} [I - ZY]^{-1}(A_P + ZC^*(D+D^*)^{-1}C)[I - ZY]) & B(D+D^*)^{-1/2} \\ \hline -(D+D^*)^{-1/2}C(I - ZY) & D \end{array}\right)$$

The following Lemma, that is proved in the standard way, states that the P-characteristic of a positive real system is stable and minimal.

LEMMA 4.3. *The P-characteristic of a minimal positive-real system is stable and minimal. The P-characteristics of two equivalent systems are equivalent.*

We now construct the inverse map to the P-characteristic. In order to do this we first have to investigate the solutions of the Lyapunov equations of the characteristic system.

LEMMA 4.4. *Let* $\left(\begin{array}{c|c} A & B \\ \hline C & D \end{array}\right)$ *be a minimal positive-real system and let*

$$\left(\begin{array}{c|c} \mathcal{A} & \mathcal{B} \\ \hline \mathcal{C} & \mathcal{D} \end{array}\right) := \left(\begin{array}{c|c} A_P + B(D+D^*)^{-1}B^*Y & B(D+D^*)^{-1/2} \\ \hline -(D+D^*)^{-1/2}C(I - ZY) & D \end{array}\right)$$

be its P-characteristic system, with Y and Z the stabilizing solutions to the respective Riccati equations. Then the Lyapunov equations

$$\mathcal{A}P + P\mathcal{A}^* = -\mathcal{B}\mathcal{B}^*$$

$$\mathcal{A}^*Q + Q\mathcal{A} = -\mathcal{C}^*\mathcal{C}$$

have solutions given by

$$P = (I - ZY)^{-1}Z = Z(I - YZ)^{-1}$$

$$Q = Y(I - ZY) = (I - YZ)Y.$$

Proof. We first show that $\mathcal{A}P + P\mathcal{A}^* = -\mathcal{B}\mathcal{B}^*$. To do this note that $I - ZY$ is invertible and consider,

$$(I - ZY)(\mathcal{A}P + P\mathcal{A}^*)(I - YZ)$$

$$= (I - ZY)\mathcal{A}Z + Z\mathcal{A}^*(I - YZ)$$

$$= (I-ZY)[A_P+B(D+D^*)^{-1}B^*Y]Z+Z[A_P+B(D+D^*)^{-1}B^*Y]^*(I-YZ)$$

$$= A_P Z + Z A_P^* - Z(Y A_P + A_P^* Y)Z$$

$$+(I - ZY)B(D + D^*)^{-1}B^*YZ + ZYB(D + D^*)^{-1}B^*(I - YZ)$$

$$= -ZC^*(D + D^*)^{-1}CZ - B(D + D^*)^{-1}B^*$$

$$-Z[-YB(D + D^*)^{-1}B^*Y - C^*(D + D^*)^{-1}C]Z$$

$$+(I - ZY)B(D + D^*)^{-1}B^*YZ + ZYB(D + D^*)^{-1}B^*(I - YZ)$$

$$= -B(D + D^*)^{-1}B^* + ZYB(D + D^*)^{-1}B^*YZ$$

$$+(I - ZY)B(D + D^*)^{-1}B^*YZ + ZYB(D + D^*)^{-1}B^*(I - YZ)$$

$$= -(I - ZY)B(D + D^*)^{-1}B^*(I - YZ)$$

$$= -(I - ZY)\mathcal{B}\mathcal{B}^*(I - YZ).$$

Let now $Q = Y - YZY$. We are now going to show that

$$\mathcal{A}^*Q + Q\mathcal{A} = -\mathcal{C}^*\mathcal{C}.$$

We consider

$$\mathcal{A}^*Q + Q\mathcal{A}$$

$$= [A_P+B(D+D^*)^{-1}B^*Y]^*Y(I-ZY)+(I-YZ)Y[A_P+B(D+D^*)^{-1}B^*Y]$$

$$= [A_P^*Y + YB(D + D^*)^{-1}B^*Y](I - ZY) + (I - YZ)[YA_P + YB(D + D^*)^{-1} B^*Y],$$

using the positive real Riccati equation this gives,

$$= [-YA_P - C^*(D + D^*)^{-1}C](I-ZY) + (I - YZ)[-A_P^*Y - C^*(D + D^*)^{-1}C]$$

$$= -C^*(D+D^*)^{-1}C(I-ZY)-(I-YZ)C^*(D+D^*)^{-1}C$$

$$-YA_P - A_P^*Y + Y(A_PZ + ZA_P^*)Y$$

$$= -C^*(D+D^*)^{-1}(I-ZY)-(I-YZ)C^*(D+D^*)^{-1}C$$

$$+YB(D+D^*)^{-1}B^*Y + C^*(D+D^*)^{-1}C$$

$$+Y[-ZC^*(D+D^*)^{-1}CZ - B(D+D^*)^{-1}B^*]Y$$

$$= -(I-YZ)C^*(D+D^*)^{-1}C(I-ZY)$$

$$= -\mathcal{C}^*\mathcal{C}.$$

□

A consequence of the previous Lemma is that the image of χ_P is not the whole set $C_n^{m,m}$, but the subset $UC_{n,P}^{m,m}$. A system $(\mathcal{A},\mathcal{B},\mathcal{C},\mathcal{D}) \in C_n^{m,m}$ is in $UC_{n,P}^{m,m}$ if

1. $\lambda_{max}(PQ) < 1$, where P, Q are the positive definite solutions to

$$\mathcal{A}P + P\mathcal{A}^* = -\mathcal{B}\mathcal{B}^*, \quad \mathcal{A}^*Q + Q\mathcal{A} = -\mathcal{C}^*\mathcal{C},$$

2. $\mathcal{D} + \mathcal{D}^* > 0$.

We can now show that the P-characteristic of a positive real system is minimal and in $UC_{n,P}^{m,m}$.

LEMMA 4.5. *We have*

$$\chi_B(P_n^m) \subseteq UC_{n,P}^{m,m}.$$

Proof. The proof is analogous to the proof of Lemma 3.5. □

In the next definition we are going to define the inverse P-characteristic map $I\chi_P : UC_{n,P}^{m,m} \to P_n^m$. Analogously to the case of minimal and bounded real systems $I\chi_P$ will turn out to be the inverse of the P-characteristic map $\chi_P : P_n^m \to UC_{n,P}^{m,m}$.

DEFINITION 4.6. *Let $(\mathcal{A},\mathcal{B},\mathcal{C},\mathcal{D}) \in UC_{n,P}^{m,m}$ and let P, Q be the solutions to the Lyapunov equations*

$$\mathcal{A}P + P\mathcal{A}^* = -\mathcal{B}\mathcal{B}^*, \quad \mathcal{A}^*Q + Q\mathcal{A} = -\mathcal{C}^*\mathcal{C}.$$

Then

$$I\chi_P\left(\left(\begin{array}{c|c} \mathcal{A} & \mathcal{B} \\ \hline \mathcal{C} & \mathcal{D}\end{array}\right)\right) := \left(\begin{array}{c|c} \mathcal{A} - \mathcal{B}(\mathcal{B}^*Q + \mathcal{C})(I-PQ)^{-1} & \mathcal{B}(\mathcal{D}^* + \mathcal{D})^{1/2} \\ \hline -(\mathcal{D}+\mathcal{D}^*)^{1/2}\mathcal{C}(I-PQ)^{-1} & \mathcal{D}\end{array}\right)$$

is called the inverse P-characteristic system.

LEMMA 4.7. *Let* $(\mathcal{A},\mathcal{B},\mathcal{C},\mathcal{D}) \in UC^{m,m}_{n,P}$ *and let* P *and* Q *be such that*

$$\mathcal{A}P + P\mathcal{A}^* = -\mathcal{B}\mathcal{B}^*, \quad \mathcal{A}^*Q + Q\mathcal{A} = -\mathcal{C}^*\mathcal{C}.$$

Then,

$$[\mathcal{A}-\mathcal{B}(\mathcal{B}^*Q+\mathcal{C})(I-PQ)^{-1}][I-PQ] = [I-PQ][\mathcal{A}-(I-PQ)^{-1}(P\mathcal{C}^*+\mathcal{B})\mathcal{C}].$$

Proof. We have

$$[\mathcal{A}-\mathcal{B}(\mathcal{B}^*Q+\mathcal{C})(I-PQ)^{-1}][I-PQ]$$

$$= \mathcal{A}(I-PQ)-\mathcal{B}(\mathcal{B}^*Q+\mathcal{C}) = \mathcal{A}-\mathcal{A}PQ-\mathcal{B}\mathcal{B}^*Q-\mathcal{B}\mathcal{C} = \mathcal{A}-[\mathcal{A}P+\mathcal{B}\mathcal{B}^*]Q-\mathcal{B}\mathcal{C}$$

$$= \mathcal{A}-[-P\mathcal{A}^*]Q-\mathcal{B}\mathcal{C} = \mathcal{A}-PQ\mathcal{A}-P\mathcal{C}^*\mathcal{C}-\mathcal{B}\mathcal{C}$$

$$= [I-PQ][\mathcal{A}-(I-PQ)^{-1}(P\mathcal{C}^*+\mathcal{B})\mathcal{C}].$$

□

In order to be able to show that $I\chi_P$ maps a stable system in $UC^{m,m}_{n,P}$ to a positive real system was need the following Lemma.

LEMMA 4.8. *Let* $(\mathcal{A},\mathcal{B},\mathcal{C},\mathcal{D}) \in UC^{m,m}_{n,P}$. *Let* P, Q *be the positive definite solutions to the Lyapunov equations*

$$\mathcal{A}P + P\mathcal{A}^* + \mathcal{B}\mathcal{B}^* = 0, \quad \mathcal{A}^*Q + Q\mathcal{A} + \mathcal{C}^*\mathcal{C} = 0.$$

Let

$$\left(\begin{array}{c|c} A & B \\ \hline C & D \end{array}\right) := I\chi_P\left(\left(\begin{array}{c|c} \mathcal{A} & \mathcal{B} \\ \hline \mathcal{C} & \mathcal{D} \end{array}\right)\right).$$

Then,

$$0 = A_P^*Y + YA_P + YB(D+D^*)^{-1}B^*Y + C^*(D+D^*)^{-1}C,$$

$$0 = A_PZ + ZA_P^* + ZC^*(D+D^*)^{-1}CZ + B(D+D^*)^{-1}B^*,$$

with

$$Y = Q(I-PQ)^{-1} = (I-QP)^{-1}Q,$$

$$Z = (I-PQ)P = P(I-QP).$$

Moreover, Y *and* Z *are the stabilizing solutions to these two positive real Riccati equations.*

Proof. The proof is completely analogous to the proof of Lemma 3.8. □

We now have that $I\chi_P$ maps $UC^{m,m}_{n,P}$ into P^m_n.

LEMMA 4.9. *We have*

$$I\chi_P(UC^{m,m}_{n,B}) \subseteq P^m_n.$$

Proof. The proof is analogous to the proof of Lemma 3.9 . □

We are now in a position to show that $\chi_P : P^m_n \to UC^{m,m}_{n,P}$ is a bijection.

THEOREM 4.10. *The map*

$$\chi_P : P^{m,m}_n \to UC^{m,m}_{n,P}$$

is a bijection that preserves system equivalence.

The inverse map $\chi_P^{-1} : UC^{m,m}_{n,P} \to P^m_n$ *is given by* $\chi_P^{-1} = I\chi_P$.

Proof. The proof is a straightforward verification and analogous to the bounded real case. □

Balancing for positive real systems has been introduced by Desai and Pal [7] (see also [14], [11]).

DEFINITION 4.11. *A system* $(A, B, C, D) \in P^m_n$ *is called* positive real balanced *if*

$$A_P^*\Sigma + \Sigma A_P + \Sigma B(D + D^*)^{-1}B^*\Sigma + C^*(D + D^*)^{-1}C = 0,$$

$$A_P\Sigma + \Sigma A_P^* + \Sigma C^*(D + D^*)^{-1}C\Sigma + B(D + D^*)^{-1}B^* = 0,$$

for

$$\Sigma = diag(\sigma_1 I_{n_1}, \sigma_2 I_{n_2}, \ldots, \sigma_k I_{n_k}), \quad \sigma_1 > \sigma_2 > \cdots > \sigma_k > 0,$$

and Σ *is the stabilizing solution to both equations. The matrix* Σ *is called the* positive real grammian *of the system.*

In Ober [21] the following canonical form for SISO positive real systems was given.

THEOREM 4.12. *The following two statements are equivalent:*

(i) $g(s)$ *is the transfer function of a positive real system over* $\Re$ *of McMillan degree* n.

(ii) $g(s)$ *has a standard* n*-dimensional realization* (A, b, c, d) *given by a standard set of parameters such that*

$$a_{ij} = \frac{-b_i b_j (1 - s_i\sigma_i)(1 - s_j\sigma_j)}{2d(s_i s_j \sigma_i + \sigma_j)},$$

with $d > 0$, $\sigma_1 < 1$.

Moreover, the realization given in (ii) is positive real balanced with positive real grammian $\Sigma = diag(\sigma_1 I_{n_1}, \sigma_2 I_{n_2}, \ldots, \sigma_k I_{n_k})$. *The map which assigns to each positive real system the realization in (ii) is a canonical form.*

As in the case of minimal systems we now introduce a slightly modified characteristic map. Define for a system $(A, B, C, D) \in P_n^m$ the modified P-characteristic map

$$\tilde{\chi}_P\left(\left(\begin{array}{c|c} A & B \\ \hline C & D \end{array}\right)\right)$$

$$= \left(\begin{array}{c|c} T^{1/4}(A_P + B(D+D^*)^{-1}B^*Y)T^{-1/4} & T^{1/4}B(D+D^*)^{-1/2} \\ \hline -(D+D^*)^{-1/2}C(I-ZY)T^{-1/4} & D \end{array}\right),$$

where $T := (I - ZY)^*(I - ZY)$ and Z, Y are the stabilizing solutions to the two positive real Riccati equations. We have the following corollary.

COROLLARY 4.13. *The map* $\tilde{\chi}_P : P_n^m \to UC_{n,P}^{m,m}$ *is a bijection with the following properties,*

1. $(A, B, C, D) \in P_n^{m,m}$ *is positive real balanced with positive real grammian* Σ *if and only if* $\tilde{\chi}_P((A, B, C, D))$ *is Lyapunov balanced with Lyapunov grammian* Σ.
2. $(A, b, c, d) \in P_n^1$ *is in positive real balanced canonical form of Theorem 4.12 if and only if* $\tilde{\chi}_P((A, b, c, d))$ *is in Lyapunov balanced canonical form of Theorem 1.1.*

5. Antistable systems. The last class of systems that we will consider is the class of antistable functions. We call a system antistable whose eigenvalues are all in the open left half plane. In this section we are going to study a map from the set of antistable systems of fixed McMillan degree to the set of stable systems of McMillan degree n. There are of course a number of obvious such maps. This map here however appears to be different.

DEFINITION 5.1. *Let* $\left(\begin{array}{c|c} A & B \\ \hline C & D \end{array}\right)$ *be a minimal antistable system. Let* Y *be the stabilizing solution of the Riccati equation*

$$A^*Y + YA - YBB^*Y = 0,$$

i.e. $A - BB^*Y$ *is stable, and let* Z *be the stabilizing solution to the Riccati equation*

$$AZ + ZA^* - ZC^*CZ = 0,$$

i.e. $A - ZC^*C$ *is stable.*

Then the system

$$\chi_S\left(\left(\begin{array}{c|c} A & B \\ \hline C & D \end{array}\right)\right) := \left(\begin{array}{c|c} A - BB^*Y & B \\ \hline CZY & D \end{array}\right)$$

is called the S-characteristic *of the system* $\left(\begin{array}{c|c} A & B \\ \hline C & D \end{array}\right)$.

These following relations are the relations that are equivalent to the Bucy relations for the case of minimal systems.

LEMMA 5.2.

Let $\left(\begin{array}{c|c} A & B \\ \hline C & D \end{array}\right)$ *be a minimal antistable system. Let* Y *be a solution of the Riccati equation*

$$0 = A * Y + YA - YBB^*Y,$$

and let Z *be a solution of the Riccati equation*

$$0 = AZ + ZA^* - ZC^*CZ,$$

then

$$ZY(A - BB^*Y) = (A - ZC^*C)ZY.$$

Proof. Consider the two Riccati equations,

$$0 = A^*Y + YA - YBB^*Y$$

and

$$0 = AZ + ZA^* - ZC^*CZ.$$

Multiplying the first equation on the left by Z and the second equation on the right by Y, equating both equations, we obtain

$$ZA^*Y + ZYA - ZYBB^*Y = AZY + ZA^*Y - ZC^*CZY.$$

Canceling the term ZA^*Y from either side and collecting terms, we obtain

$$ZY(A - BB^*Y) = (A - ZC^*C)ZY.$$

□

As a consequence of these Bucy type relationships we can rewrite the characteristic of a system as follows,

$$\chi_P\left(\left(\begin{array}{c|c} A & B \\ \hline C & D \end{array}\right)\right) := \left(\begin{array}{c|c} [ZY]^{-1}(A - ZC^*C)[ZY] & B \\ \hline CZY & D \end{array}\right).$$

The following Lemma shows that the S-characteristic maps antistable systems to stable minimal systems. The proof is by now standard.

LEMMA 5.3. *The S-characteristic of a minimal anti-stable system is stable and minimal. The S-characteristics of two equivalent systems are equivalent.*

In the following definition we are going to introduce the candidate map for the inverse of the characteristic map. We denote by $AS_n^{p,m} \subseteq L_n^{p,m}$ the subset of antistable systems.

DEFINITION 5.4. *Let* $(\mathcal{A}, \mathcal{B}, \mathcal{C}, \mathcal{D}) \in C_n^{p,m}$ *and let* P, Q *be the solutions to the Lyapunov equations*

$$\mathcal{A}P + P\mathcal{A}^* = -\mathcal{B}\mathcal{B}^*, \quad \mathcal{A}^*Q + Q\mathcal{A} = -\mathcal{C}^*\mathcal{C}.$$

Then

$$I\chi_S\left(\left(\begin{array}{c|c} \mathcal{A} & \mathcal{B} \\ \hline \mathcal{C} & \mathcal{D} \end{array}\right)\right) := \left(\begin{array}{c|c} \mathcal{A} + \mathcal{B}\mathcal{B}^* P^{-1} & \mathcal{B} \\ \hline \mathcal{C}(PQ)^{-1} & \mathcal{D} \end{array}\right)$$

is called the inverse S-characteristic system.

To show that $I\chi_S$ maps a stable minimal system to an antistable minimal system we need the following Lemma.

LEMMA 5.5. *Let* $(\mathcal{A}, \mathcal{B}, \mathcal{C}, \mathcal{D}) \in C_n^{p,m}$ *and let* P *and* Q *be such that*

$$\mathcal{A}P + P\mathcal{A}^* = -\mathcal{B}\mathcal{B}^*, \quad \mathcal{A}^*Q + Q\mathcal{A} = -\mathcal{C}^*\mathcal{C}.$$

Then

$$[\mathcal{A} + \mathcal{B}\mathcal{B}^* P^{-1}]PQ = PQ[\mathcal{A} + Q^{-1}\mathcal{C}^*\mathcal{C}]$$

Proof. We have

$$[\mathcal{A} + \mathcal{B}\mathcal{B}^* P^{-1}]PQ = \mathcal{A}PQ + \mathcal{B}\mathcal{B}^*Q$$

$$= [\mathcal{A}P + \mathcal{B}\mathcal{B}^*]Q = [-P\mathcal{A}^*]Q = -P[-\mathcal{C}^*\mathcal{C} - Q\mathcal{A}]$$

$$= PQ\mathcal{A} + P\mathcal{C}^*\mathcal{C} = PQ[\mathcal{A} + Q^{-1}\mathcal{C}^*\mathcal{C}].$$

□

LEMMA 5.6. *We have*

$$I\chi_S(C_n^{p,m}) \subseteq AS_n^{p,m}.$$

Proof. Let $(\mathcal{A}, \mathcal{B}, \mathcal{C}, \mathcal{D}) \in C_n^{p,m}$. The minimality of $(A, B, C, D) = I\chi_S((\mathcal{A}, \mathcal{B}, \mathcal{C}, \mathcal{D}))$ follows in the standard way. To show that A is antistable let P be the positive definite solution to the Lyapunov equation,

$$\mathcal{A}P + P\mathcal{A}^* = -\mathcal{B}\mathcal{B}^*.$$

Now consider,

$$AP + PA^* - BB^* = (\mathcal{A} + \mathcal{B}\mathcal{B}^* P^{-1})P + P(\mathcal{A} + \mathcal{B}\mathcal{B}^* P^{-1})^* - \mathcal{B}\mathcal{B}^*$$

$$= \mathcal{A}P + P\mathcal{A}^* + \mathcal{B}\mathcal{B}^* = 0,$$

which implies that A is antistable. Hence $(A, B, C, D) \in AS_n^{p,m}$. □

In order to prove the main result of this section we will need to again establish connections between solutions to Lyapunov equations and solutions to Riccati equations.

LEMMA 5.7. *Let* $\left(\begin{array}{c|c} A & B \\ \hline C & D \end{array}\right)$ *be a minimal antistable system and let*

$$\left(\begin{array}{c|c} \mathcal{A} & \mathcal{B} \\ \hline \mathcal{C} & \mathcal{D} \end{array}\right) := \left(\begin{array}{c|c} A - BB^*Y & B \\ \hline CZY & D \end{array}\right)$$

be its S*-characteristic system, with* Y *and* Z *the stabilizing solutions to the respective Riccati equations. Then the Lyapunov equations*

$$\mathcal{A}P + P\mathcal{A}^* = -\mathcal{B}\mathcal{B}^*$$

$$\mathcal{A}^*Q + Q\mathcal{A} = -\mathcal{C}^*\mathcal{C}$$

have solutions given by

$$P = Y^{-1}$$

$$Q = YZY.$$

Proof. We first show that $\mathcal{A}P + P\mathcal{A}^* = -\mathcal{B}\mathcal{B}^*$. To do this consider

$$Y(\mathcal{A}P + P\mathcal{A}^*)Y = Y\mathcal{A} + \mathcal{A}^*Y = A^*Y + YA - YBB^*Y - YBB^*Y,$$

using the Riccati equation this gives,

$$= YBB^*Y - 2YBB^*Y = -Y\mathcal{B}\mathcal{B}Y,$$

which implies the claim. Now, let $Q = YZY$ and consider,

$$\mathcal{A}^*Q + Q\mathcal{A} = (A - BB^*Y)^*YZY + YZY(A - BB^*Y),$$

using the Bucy type relations, this gives

$$= [(ZY)^{-1}(A - ZC^*C)ZY]^*YZY + YZY(ZY)^{-1}(A - ZC^*C)ZY$$

$$= YZ(A - ZC^*C)^*Y + Y(A - ZC^*C)ZY$$

$$= Y(AZ + ZA^*)Y - 2YZC^*CZY,$$

using the Riccati equation in Z, we obtain,

$$= YZC^*CZY - 2YZC^*CY = -YZC^*CZY = -\mathcal{C}^*\mathcal{C},$$

which implies the result. □

LEMMA 5.8. *Let $(\mathcal{A},\mathcal{B},\mathcal{C},\mathcal{D}) \in C_n^{p,m}$. Let P, Q be the positive definite solutions to the Lyapunov equations*

$$\mathcal{A}P + P\mathcal{A}^* + \mathcal{B}\mathcal{B}^* = 0, \quad \mathcal{A}^*Q + Q\mathcal{A} + \mathcal{C}^*\mathcal{C} = 0.$$

Let

$$\left(\begin{array}{c|c} A & B \\ \hline C & D \end{array}\right) := I\chi_P\left(\left(\begin{array}{c|c} \mathcal{A} & \mathcal{B} \\ \hline \mathcal{C} & \mathcal{D} \end{array}\right)\right).$$

Then,

$$0 = A^*Y + YA - YBB^*Y,$$

$$0 = A_P Z + ZA_P^* - ZC^*CZ,$$

with

$$Y = P^{-1},$$

$$Z = PQP.$$

Moreover, Y and Z are the stabilizing solutions to these two Riccati equations.

Proof. Since,

$$QP[A^*Y + YA - YBB^*Y]PQ$$

$$= QP[(\mathcal{A}+\mathcal{B}\mathcal{B}^*P^{-1})^*P^{-1} + P^{-1}(\mathcal{A}+\mathcal{B}\mathcal{B}^*P^{-1}) - P^{-1}\mathcal{B}\mathcal{B}^*P^{-1}]PQ$$

$$= QP\mathcal{A}^*Q + Q\mathcal{B}\mathcal{B}^*Q + Q\mathcal{A}PQ + Q\mathcal{B}\mathcal{B}^*Q - Q\mathcal{B}\mathcal{B}^*Q$$

$$= Q(P\mathcal{A}^* + \mathcal{A}P + \mathcal{B}\mathcal{B}^*)Q$$

$$= 0,$$

we have verified the first identity. Now with $Z = PQP$ we have

$$AZ + ZA^* - ZC^*CZ$$

$$= (\mathcal{A}+\mathcal{B}\mathcal{B}^* P^{-1})PQP+PQP(\mathcal{A}+\mathcal{B}\mathcal{B}^* P^{-1})^* - PQP(PQ)^{-*}\mathcal{C}^*\mathcal{C}(PQ)^{-1}PQP$$

$$= \mathcal{A}PQP + \mathcal{B}\mathcal{B}^*QP + PQP\mathcal{A}^* + PQ\mathcal{B}\mathcal{B}^* - P\mathcal{C}^*\mathcal{C}P$$

$$= (\mathcal{A}P + \mathcal{B}\mathcal{B}^*)QP - P\mathcal{C}^*\mathcal{C}P + PQ(P\mathcal{A}^* + \mathcal{B}\mathcal{B}^*)$$

$$= -P\mathcal{A}^*QP - P\mathcal{C}^*\mathcal{C}P + PQ(P\mathcal{A}^* + \mathcal{B}\mathcal{B}^*)$$

$$= -P(\mathcal{A}^*Q + \mathcal{C}^*\mathcal{C})P + PQ(P\mathcal{A}^* + \mathcal{B}\mathcal{B}^*)$$

$$= P(Q\mathcal{A})P + PQ(P\mathcal{A}^* + \mathcal{B}\mathcal{B}^*)$$

$$= PQ(\mathcal{A}P + P\mathcal{A}^* + \mathcal{B}\mathcal{B}^*)$$

$$= 0,$$

which shows the second identity. Since

$$A - BB^*Y = \mathcal{A} + \mathcal{B}\mathcal{B}^* P^{-1} - \mathcal{B}\mathcal{B}^* P^{-1} = \mathcal{A},$$

which is stable and

$$A - ZC^*C = \mathcal{A} + \mathcal{B}\mathcal{B}^* P^{-1} - PQP(PQ)^{-*}\mathcal{C}^*\mathcal{C}(PQ)^{-1}$$

$$= \mathcal{A} + \mathcal{B}\mathcal{B}^* P^{-1} - P\mathcal{C}^*\mathcal{C}(PQ)^{-1}$$

$$= PQ[\mathcal{A} + Q^{-1}\mathcal{C}^*\mathcal{C}](PQ)^{-1} - P\mathcal{C}^*\mathcal{C}(PQ)^{-1}$$

$$= (PQ)\mathcal{A}(PQ)^{-1},$$

is stable, where we have used Lemma 5.5, we have shown that Z, Y are the stabilizing solutions to the Riccati equations. □

We now state the main theorem of this section which shows that the S-characteristic map is a bijection.

THEOREM 5.9. *The map*

$$\chi_S : AS_n^{p,m} \to \mathcal{C}_n^{p,m}$$

is a bijection that preserves system equivalence. The inverse map χ_S^{-1} : $\mathcal{C}_n^{p,m} \to AS_n^{p,m}$ *is given by* $\chi_S^{-1} = I\chi_S$.

Proof. The proof is analogous to the proof of Theorem 3.10. □

Following the examples of the previous sections we now introduce a balancing scheme for antistable systems.

DEFINITION 5.10. *A system* $(A, B, C, D) \in AS_n^{p,m}$ *is called* anti-stable balanced *if*

$$A^*\Sigma + \Sigma A - \Sigma BB^*\Sigma = 0,$$

$$A\Sigma + \Sigma A^* - \Sigma C^*C\Sigma = 0,$$

for

$$\Sigma = diag(\sigma_1 I_{n_1}, \sigma_2 I_{n_2}, \ldots, \sigma_k I_{n_k}), \quad \sigma_1 > \sigma_2 > \cdots > \sigma_k > 0,$$

and Σ *is the stabilizing solution to both equations. The matrix* Σ *is called the* anti-stable grammian *of the system.*

We are again going to define a modified characteristic map. Let $(A, B, C, D) \in AS_n^{p,m}$ then define

$$\tilde{\chi}_S\left(\left(\begin{array}{c|c} A & B \\ \hline C & D \end{array}\right)\right) = \left(\begin{array}{c|c} T^{1/4}(A - BB^*Y)T^{-1/4} & T^{1/4}B \\ \hline CZYT^{-1/4} & D \end{array}\right),$$

where $T := (ZY)^*(ZY)$ and Z, Y are the stabilizing solutions to the two Riccati equations. We have the following corollary.

COROLLARY 5.11. *The map* $\tilde{\chi}_S : AS_n^{p,m} \rightarrow C_n^{p,m}$ *is a bijection such that,* $(A, B, C, D) \in AS_n^{p,m}$ *is antistable balanced with anti-stable grammian* Σ *if and only if* $\tilde{\chi}_S((A, B, C, D))$ *is Lyapunov balanced with Lyapunov grammian* Σ.

6. Diffeomorphisms. In the previous sections we have studied bijections between various sets of linear systems. We are now going to show that these maps are in fact diffeomorphisms.

We will need a new notation. To indicate the subsets of strictly proper systems we append the additional subscript 0, e.g. $C_{n,0}^{p,m}$ denotes the strictly proper systems in $C_n^{p,m}$. To indicate the subsets with identity feedthrough term we append the subscript I, e.g. $UC_{n,I}^{m,m}$ denotes the subset of $UC_{n,P}^{m,m}$ with $D = I$.

These sets can be endowed with a topology by embedding them in a natural way in Euclidean space. If we denote by $\sim$ the equivalence relation given by system equivalence then we consider the quotient spaces $L_{n,0}^{p,m} / \sim$, $C_{n,0}^{p,m} / \sim$, ..., to be endowed with the quotient topology.

The key to our way of proceeding is a result by Delchamps [4], [5],[6] that states that stabilizing solutions to Riccati equations are differentiable functions. Similarly, the positive definite solutions to Lyapunov equations are differentiable function of the system matrices. This implies that the bijections that we have constructed are in fact diffeomorphisms. In the same series of papers Delchamps also showed that $B_{n,0}^{p,m}$ and $C_{n,0}^{p,m}$ are diffeomorphic.

From this discussion we immediately have the following result.

THEOREM 6.1. *We have that*

$$C_{n,0}^{p,m}/\sim,\ B_{n,0}^{p,m}/\sim,\ L_{n,0}^{p,m}/\sim,\ UC_{n,0}^{p,m}/\sim,\ AS_{p,0}^{p,m}/\sim$$

are diffeomorphic. Moreover,

$$C_{n,0}^{m,m}/\sim,\ B_{n,0}^{m,m}/\sim,\ L_{n,0}^{m,m}/\sim,\ AS_{p,0}^{m,m}/\sim,\ UC_{n,0}^{m,m}/\sim, UC_{n,I}^{m,m}/\sim,$$

$$P_{n,I}^{m}/\sim$$

are diffeomorphic.

REFERENCES

[1] B. D. O. Anderson. On the computation of the Cauchy index. *Quarterly of applied mathematics*, pages 577–582, 1972.

[2] R. W. Brockett. Some geometric questions in the theory of linear systems. *IEEE Transactions on Automatic Control*, 21:449–454, 1976.

[3] R. Bucy. The Riccati equation and its bounds. *Journal Computer Systems Science*, 6:343–353, 1972.

[4] D. Delchamps. *A Note on the Analyticity of the Riccati Metric*, volume 18, pages 37–42. AMS, Providence, RI, 1980. C.I. Byrnes and C.F. Martin, eds.

[5] D. Delchamps. Analytic stabilization and the algebraic Riccati equation. In *Proceedings 22nd Conference on Decision and Control*, San Antonio, Texas, USA, December 1983.

[6] D. Delchamps. Analytic feedback control and the algebraic Riccati equation. *IEEE Transactions on Automatic Control*, 29(11), 1984.

[7] U. B. Desai and D. Pal. A transformation approach to stochastic model reduction. *IEEE Transactions on Automatic Control*, 29:1097–1100, 1984.

[8] P. Fuhrmann and R. Ober. A functional approach to LQG-balanced realizations. International Journal of Control, 57: 627-741, 1993.

[9] K. Glover. Some geometrical properties of linear systems with implications in identification. In *Proceedings IFAC World Congress, Boston*, 1975.

[10] K. Glover and D. McFarlane. Robust stabilization of normalized coprime factors: An explicit H^∞ solution. In *Proceedings Automatic Control Conference, Atlanta, Georgia, USA*, 1988.

[11] M. Green. Balanced stochastic realizations. *Linear Algebra and Its Applications*, 98:211–247, 1988.

[12] B. Hanzon. *Identifiability, Recursive Identification and Spaces of Linear Systems.* PhD thesis, Erasmus University Rotterdam, 1986.

[13] B. Hanzon. On the differentiable manifold of fixed order stable linear systems. *Systems and Control Letters*, 13:345–352, 1989.

[14] P. Harshavardhana, E. Jonckheere, and L. Silverman. Stochastic balancing and approximation - stability and minimality. *IEEE Transactions on Automatic Control*, 29(8):744–746, 1984.

[15] U. Helmke. A global parametrization of asymptotically stable linear systems. *Systems and Control Letters*, 13:383–389, 1989.

[16] E. A. Jonckheere and L. M. Silverman. A new set of invariants for linear systems - application to reduced order compensator design. *IEEE Transactions on Automatic Control*, 28:953–964, 1983.

[17] B. C. Moore. Principal component analysis in linear systems: controllability, observability and model reduction. *IEEE Transactions on Automatic Control*, 26:17–32, 1981.

[18] R. Ober. Balanced realizations: canonical form, parametrization, model reduction. *International Journal of Control*, 46(2):643–670, 1987.

[19] R. Ober. Topology of the set of asymptotically stable systems. *International Journal of Control*, 46(1):263–280, 1987.

[20] R. Ober. Connectivity properties of classes of linear systems. *International Journal of Control*, 50:2049–2073, 1989.

[21] R. Ober. Balanced parametrization of classes of linear systems. *SIAM Journal on Control and Optimization*, 29:1251–1287, 1991.

[22] R. Ober and D. McFarlane. Balanced canonical forms for minimal systems: a normalized coprime factor approach. *Linear Algebra and its Applications, Special Issue on Linear Systems and Control*, 122–124:23–64, 1989.

[23] P. C. Opdenacker and E. A. Jonckheere. A contraction mapping preserving balanced reduction scheme and its infinity norm error bounds. *IEEE Transactions on Circuits and Systems*, 35:184–189, 1988.

[24] J. C. Willems. Least squares stationary optimal control and the algebraic Riccati equation. *IEEE Transactions on Automatic Control*, 16:621–634, 1971.

TRANSFER FUNCTION APPROACH TO DISTURBANCE DECOUPLING PROBLEM

MAREK RAKOWSKI*

Abstract. We give a necessary and sufficient condition for existence of a controller which decouples the disturbance signal. The condition is based on state space computations. If it is satisfied, we parametrize the set of all disturbance decoupling controllers.

Key words. disturbance decoupling problem, generalized inversion

AMS(MOS) subject classifications. 16G70, 93B20, 93C45

1. Introduction. Consider the system

$$\begin{cases} x'(t) &= Ax(t) + Bu_c(t) + B_d u_d(t) \\ y(t) &= Cx(t) \\ y_m(t) &= C_m x(t) \end{cases} \tag{1.1}$$

where A, B, B_d, C, and C_m are matrices, u_d is the disturbance, y the output, and y_m the measurement output. The *disturbance decoupling problem* is to find a controller (or a compensator)

$$\begin{cases} x_c'(t) &= A_c x_c(t) + B_c y_m(t) \\ u_c(t) &= C_c x_c(t) + D_c y_m(t) \end{cases} \tag{1.2}$$

so that the transfer function from u_d to y is identically zero.

Suppose a controller (1.2) is given. Equations (1.1) and (1.2) can be combined into

$$\begin{cases} \begin{bmatrix} x'(t) \\ x_c'(t) \end{bmatrix} = \begin{bmatrix} A + BD_cC_m & BC_c \\ B_cC_m & A_c \end{bmatrix} \begin{bmatrix} x(t) \\ x_c(t) \end{bmatrix} + \begin{bmatrix} B_d \\ 0 \end{bmatrix} u_d(t) \\ y(t) = \begin{bmatrix} C & 0 \end{bmatrix} \begin{bmatrix} x(t) \\ x_c(t) \end{bmatrix}. \end{cases} \tag{1.3}$$

Thus, if the disturbance decoupling problem has a solution, $\operatorname{im}[\,B_d^T \quad 0\,]^T$ is contained in a subspace V such that V is invariant under

$$\begin{bmatrix} A + BD_cC_m & BC_c \\ B_cC_m & A_c \end{bmatrix} \tag{1.4}$$

and $[\ C \quad 0\ ]V = (0)$. Partition the state space of the system (1.3) as $X\dot{+}X_c$ where X and X_c are the state spaces of the systems (1.1) and (1.2), and let π be the projection onto X along X_c. Let

$$V_i = \pi(V \cap (X\dot{+}(0))),$$

* Department of Mathematics, The Ohio State University, 231 West 18th Avenue, Columbus, OH 43210.

and let

$$V_p = \pi V.$$

Then [29] $\operatorname{im} B_d \subset V_i$, $V_p \subset \ker C$, and a straightforward computation shows that (V_i, V_p) is a (C_m, A, B)-pair, where in general a (C, A, B)-pair is a pair of subspaces (V_1, V_2) such that

(i) $V_1 \subset V_2$,
(ii) $A(V_1 \cap \ker C) \subset V_1$,
(iii) $A(V_2) \subset V_2 + \operatorname{im} B$.

It is well known that condition (ii) is equivalent to

(ii') $\exists G$ such that $(A + GC)V_1 \subset V_1$,

and condition (iii) is equivalent to

(iii') $\exists F$ such that $(A + FB)V_2 \subset V_2$.

A subspace V_1 which satisfies condition (ii) and (ii') is said to be (C, A)-*invariant*, and a subspace V_2 which satisfies conditions (iii) and (iii') is said to be (A, B)-*invariant*.[1]

It has been shown in [29] (see also [30]) that the preceding argument can be reversed. That is, suppose we have a system (1.1), there exists a (C_m, A, B)-pair (V_1, V_2) such that $\operatorname{im} B_d \subset V_1$ and $V_2 \subset \ker C$, and let $W = \mathbf{C}^{\dim V_2 - \dim V_1}$. Then there exist linear maps A_c, B_c, C_c and D_c, and a subspace V of $X \dot{+} W$, such that V is invariant under the operator (1.4) and $[C \quad 0]V = (0)$, and so the disturbance decoupling problem has a solution. Since the sum of (A, B)-invariant subspaces is an (A, B)-invariant subspace, there is a largest (A, B)-invariant subspace contained in $\ker C$, $V^*(\ker C)$. Similarly, since the intersection of (C, A)-invariant subspaces is a (C, A)-invariant subspace, there is a smallest (C, A)-invariant subspace containing $\operatorname{im} B_d$, $V_*(\operatorname{im} B_d)$. Thus, the disturbance decoupling problem has a solution if and only if

$$V_*(\operatorname{im} B_d) \subset V^*(\ker C). \tag{1.5}$$

If the condition (1.5) is satisfied, the minimum dimension of the state space of the controller equals [29]

$$\begin{aligned} \min\{\dim V_2 - \dim V_1 \quad : \; & \operatorname{im} B_d \subset V_1, \; V_2 \subset \ker C, \text{ and} \\ & (V_1, V_2) \text{ is a } (C_m, A, B) - \text{pair}\}. \end{aligned}$$

Below, we present a different solution of the disturbance decoupling problem which is based on the transfer functions. More specifically, we reformulate the problem in terms of the causal (that is, proper or analytic at infinity) solutions of equation

$$GXH = R, \tag{1.6}$$

[1] The name "(C, A, B)-pair" is in a sense better than the name "(C, A)-" or "(A, B)-invariant subspace" in that it does not depend on how the matrices A, B and C are labeled.

where G, H and R are rational matrix functions (Section 2). In Section 3 we discuss a general solution of equation (1.6), and in Section 4 we parametrize all causal solutions of the equation in terms of a free parameter Y which runs over all proper rational matrix functions of appropriate size. Our approach is based on the formulas for pointwise generalized inverses of rational matrix functions [23, 24].

We note that if the function G in (1.6) is identically equal to identity, the problem of finding causal solutions of the equation is called the *causal factorization problem.* Various necessary and sufficient conditions for existence of causal solutions in this case are known [19, 16, 12, 17].

2. Transfer function conditions. After taking Laplace transforms of the signals, and assuming zero initial conditions, equations (1.1) and (1.2) become

$$\left\{\begin{aligned} sx(s) &= Ax(s) + Bu_c(s) + B_d u_d(s)\\ y(s) &= Cx(s)\\ y_m(s) &= C_m x(s)\end{aligned}\right. \tag{2.1}$$

and

$$\left\{\begin{aligned} sx_c(s) &= A_c x_c(s) + B_c y_m(s)\\ u_c(s) &= C_c x_c(s) + D_c y_m(s).\end{aligned}\right. \tag{2.2}$$

From (2.1), we have the transfer functions of the open loop system

$$\begin{aligned} T_{y\leftarrow u_d}(s) &= C(s-A)^{-1}B_d, & T_{y\leftarrow u_c}(s) &= C(s-A)^{-1}B,\\ T_{y_m\leftarrow u_d}(s) &= C_m(s-A)^{-1}B_d, & T_{y_m\leftarrow u_c}(s) &= C_m(s-A)^{-1}B.\end{aligned} \tag{2.3}$$

The transfer function of the controller (2.2) is

$$T_c(s) = D_c + C_c(s-A_c)^{-1}B_c, \tag{2.4}$$

and so in the closed loop system in Fig. 2.1 we have

$$u_c = T_c(T_{y_m\leftarrow u_d}u_d + T_{y_m\leftarrow u_c}u_c). \tag{2.5}$$

Hence

$$u_c = (I - T_cT_{y_m\leftarrow u_c})^{-1}T_cT_{y_m\leftarrow u_d}u_d, \tag{2.6}$$

and

$$T_{y\leftarrow u_d} + T_{y\leftarrow u_c}(I - T_cT_{y_m\leftarrow u_c})^{-1}T_cT_{y_m\leftarrow u_d} \tag{2.7}$$

is the closed-loop transfer function from u_d to y. Thus, if there exists a controller (1.2) which solves the disturbance decoupling problem, the function $T_{y\leftarrow u_d}$ can be factored as

$$T_{y\leftarrow u_d} = T_{y\leftarrow u_c}XT_{y_m\leftarrow u_d} \tag{2.8}$$

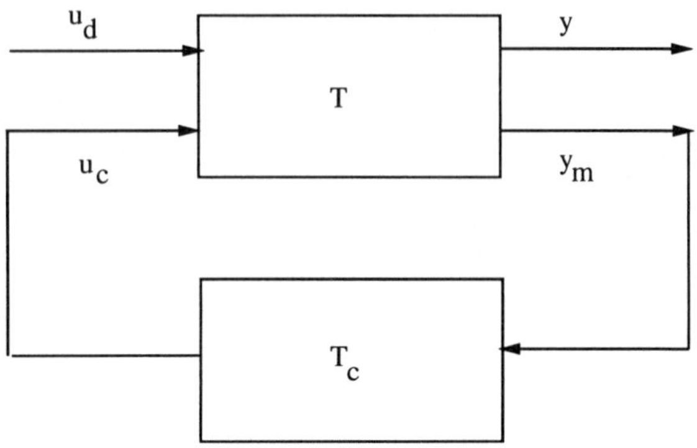

FIG. 2.1.

where

$$X = (T_c T_{y_m \leftarrow u_c} - I)^{-1} T_c \tag{2.9}$$

is proper.

Conversely, suppose we have strictly proper functions (2.3) and equation (2.8) has a proper solution X. Then we can solve equation (2.9) for T_c to obtain the controller

$$T_c = X(T_{y_m \leftarrow u_c} X - I)^{-1}. \tag{2.10}$$

Thus, proper solutions of equation (2.8) provide disturbance decoupling controllers.

In [21], a simple necessary condition for existence of proper solutions of equation (2.8) has been indicated. To simplify notation, write (2.8) as

$$R = GXH \tag{2.8'}$$

and let $G_{i,*}$ and $H_{*,j}$ denote the ith row and j^{th} column of G and H. Let $\mathcal{O}(W)$ be the order of a rational matrix function W at infinity, that is, $\mathcal{O}(0) = -\infty$ and, if $W \neq 0$, $z^{-\mathcal{O}(W)}W(z)$ is analytic and does not vanish at infinity. If equation (2.8′) has a proper solution, then

$$R_{ij} = G_{i,*} X H_{*,j} \tag{2.11}$$

and hence, for all i, j

(i) $G_{i,*} \neq 0$ and $H_{*,j} \neq 0$ whenever $R_{ij} \neq 0$,

(ii) $\mathcal{O}(R_{ij}) \leq \mathcal{O}(G_{i,*}) + \mathcal{O}(H_{*,j})$.

If the matrix G_0 (resp. H_0) formed by the leading coefficients in the Laurent expansions at infinity of the rows (resp. columns) of G (resp. H) has full row (resp. column) rank, conditions (i) and (ii) are also sufficient for existence of a proper solution of equation (2.8′). The general case can be reduced to the case when G_0 (resp. H_0) has full row (resp. column) rank by operations linear over the field of scalar rational functions.

In the disturbance decoupling problem, we are especially interested in controllers which stabilize the closed loop system in Fig. 2.1, that is, controllers for which the matrix (1.4) has all eigenvalues in the open left half-plane (we assume that realizations (1.1) and (1.2) are minimal). Such controllers are called *stabilizing compensators.* In terms of the transfer functions, stability of the system in Fig. 2.1 is equivalent [14] to the stability of the transfer functions from u_d, v_1 and v_2 to y, y_m and u_c in the following diagram.

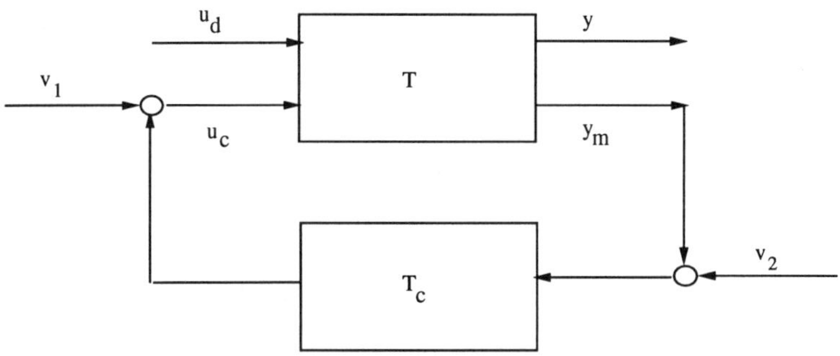

FIG. 2.2.

Suppose the plant T is *stabilizable*, that is, there exists a compensator T_c for which the system in Fig. 2.1 is stable, and let M, N, $\tilde{M}$, $\tilde{N}$, U, V, $\tilde{U}$ and $\tilde{V}$ be stable (that is, without poles in the closed right half-plane) rational matrix functions such that

$$T_{y_m \leftarrow u_c} = NM^{-1} = \tilde{M}^{-1}\tilde{N} \tag{2.12}$$

and

$$\begin{bmatrix} \tilde{U} & -\tilde{V} \\ -\tilde{N} & \tilde{M} \end{bmatrix} \begin{bmatrix} M & V \\ N & U \end{bmatrix} = I. \tag{2.13}$$

Then

$$\{(V - MQ)(U - NQ)^{-1} : Q \text{ is a stable rational matrix function}\} \tag{2.14}$$

is the set of all stabilizing compensators. The parametrization (2.14) of stabilizing compensators is called *Youla parametrization.* Another characterization of stabilizing compensators has been given in [1, 4, 5].

Below, we find all proper solutions of equation (2.8′) without addressing stability of the resulting closed loop plant. That is, we obtain all controllers which decouple the disturbance signal. The set of all stabilizing compensators which solve the disturbance decoupling problem can be obtained by intersecting the set of disturbance decoupling controllers with the set (2.14).

3. Generalized pseudoinverses. In this section we find a general solution of equation $GXH = R$ with G, H and R not necessarily proper rational matrix functions. We begin with a simplified version of the equation where all functions are constant. Consider the equation

$$AXB = C \tag{3.1}$$

where all letters represent complex matrices. If D is a matrix, by a *generalized inverse* of D we will understand any matrix $D^{\ddagger}$ such that

$$DD^{\ddagger}D = D \qquad \text{and} \qquad D^{\ddagger}DD^{\ddagger} = D^{\ddagger}. \tag{3.2}$$

There is no unique terminology concerning generalized inverses. The term generalized inverse is sometimes used to denote a linear operator which satisfies only the first equality in (3.2) (see e.g. [32]), and a matrix which satisfies both equalities (3.2) is called a (1, 2)-inverse of D in [8, 9]. In [22] the term generalized inverse has been used to denote the matrix which in addition to (3.2) satisfies equalities

$$(DD^{\ddagger})^* = DD^{\ddagger} \qquad \text{and} \qquad (D^{\ddagger}D)^* = D^{\ddagger}D. \tag{3.3}$$

This matrix is unique and is called the *Moore-Penrose inverse* of D.

Suppose equation (3.1) has a solution D. Then [22]

$$\begin{aligned} A(A^{\ddagger}CB^{\ddagger})B &= (AA^{\ddagger}A)D(BD^{\ddagger}B) \\ &= A^{\ddagger}DB^{\ddagger} \\ &= C \end{aligned} \tag{3.4}$$

and $X = A^{\ddagger}CB^{\ddagger}$ satisfies the equation. Conversely, if

$$A(A^{\ddagger}CB^{\ddagger})B = C, \tag{3.5}$$

then equation (3.1) has a solution. Thus, equation (3.1) is solvable if and only if (3.5) holds. It has been pointed out in [22] that only the first equality in (3.2) plays role in the preceding computation.

Suppose equation (3.1) is solvable and let $A^{\ddagger}$ and $B^{\ddagger}$ be generalized inverses of A and B. The particular solution $A^{\ddagger}CB^{\ddagger}$ can be used to parametrize all solutions of the equation. Indeed, a matrix Z solves (3.1) if and only if Z differs from $A^{\ddagger}CB^{\ddagger}$ by Y where Y satisfies the homogeneous equation $AXB = 0$. Suppose a matrix Y satisfies the homogeneous equation. Then

$$A^{\ddagger}(AYB)B^{\ddagger} = 0, \tag{3.6}$$

and Y can be written as

$$Y - A^{\ddagger}AYBB^{\ddagger}. \tag{3.7}$$

Conversely, for an arbitrary matrix Y of appropriate size

$$A(Y - A^{\ddagger}AYBB^{\ddagger})B = AYB - AYB = 0. \tag{3.8}$$

It follows that the general solution of equation (3.1) has the form

$$A^{\ddagger}CB^{\ddagger} + Y - A^{\ddagger}AYBB^{\ddagger}. \tag{3.9}$$

All these observations can be extended to equations involving rational matrix functions either algebraically, by considering matrices over the field $\mathcal{R}$ of scalar rational functions, or analytically, by considering pointwise generalized inverses of matrix valued functions. Adopting the notation in [6], we will denote a generalized inverse of a linear transformation $W : \mathcal{R}^n \to \mathcal{R}^m$ by $W^{\times}$. Thus,

$$WW^{\times}W = W \qquad \text{and} \qquad W^{\times}WW^{\times} = W^{\times}. \tag{3.10}$$

Let $G^{\times}$ and $H^{\times}$ be generalized inverses of G and H. Then equation (2.8′) is solvable if and only if $GG^{\times}RH^{\times}H = R$, and the general solution of the solvable equation is

$$X = G^{\times}RH^{\times} + Y - G^{\times}GYHH^{\times}. \tag{3.11}$$

We will regard an element G of $\mathcal{R}^{m\times n}$ as a (meromorphic) function from the extended complex plane $\mathbf{C}_{\infty}$ into $\mathbf{C}^{m\times n}$. The function $\overline{G}$ such that

$$\overline{G}(z)G(z) = I \qquad \text{and} \qquad G(z)\overline{G}(z) = I \tag{3.12}$$

is a pointwise inverse of G, and a function $G^{\times}$ which satisfies (3.10) is a pointwise generalized inverse of G. The functions $\overline{G}$ and $G^{\times}$ have been called a *pseudoinverse* ("fake inverse") and a *generalized pseudoinverse* in [23].

Suppose a function $G \in \mathcal{R}^{n\times n}$ is analytic and takes a nonsingular value at infinity, and let (A, B, C, D) be a realization of G. Then the pseudoinverse of G is given by the formula [31, 6]

$$[G(z)]^{-1} = D^{-1} - D^{-1}C(z - A^{\times})^{-1}BD^{-1} \tag{3.13}$$

with $A^{\times} = A - BD^{-1}C$. This formula extends to pointwise generalized inverses as follows.

Theorem 3.1 *Suppose a function $G \in \mathcal{R}^{m\times n}$ has neither a pole nor zero at infinity, and let (A, B, C, D) be a realization of G. Pick a generalized inverse $D^{\ddagger}$ of D, and let*

$$G^{\times}(z) = D^{\ddagger} - D^{\ddagger}C(z - A^{\times})^{-1}BD^{\ddagger} \tag{3.14}$$

with $A^\times = A - BD^\ddagger C$. *Then* $G^\times$ *is a generalized pseudoinverse of* G.

Proof We have

$$\begin{aligned} G^\times(z)G(z)D^\ddagger &= (D^\ddagger - D^\ddagger C(z-A^\times)^{-1}BD^\ddagger)(DD^\ddagger + C(z-A)^{-1}BD^\ddagger) \\ &= D^\ddagger + D^\ddagger C(z-A)^{-1}BD^\ddagger - D^\ddagger C(z-A^\times)^{-1}BD^\ddagger \\ &\qquad -D^\ddagger C(z-A^\times)^{-1}BD^\ddagger C(z-A)^{-1}BD^\ddagger \\ &= D^\ddagger + D^\ddagger C(z-A^\times)^{-1}[z - A^\times - z + A - BD^\ddagger C] \\ &\qquad (z-A)^{-1}BD^\ddagger \\ &= D^\ddagger. \end{aligned}$$

Hence $G^\times G G^\times = G^\times$ and the second equality in (3.10) holds. Also, $GG^\times GG^\times = GG^\times$ and multiplication by $GG^\times$ is a projection onto $\mathrm{im}\,(GG^\times)$. Clearly,

$$\mathrm{rank}_{\mathcal{R}}(GG^\times) \leq \mathrm{rank}_{\mathcal{R}}(G).$$

Since G has neither a zero nor pole at infinity and $DD^\ddagger D = D$,

$$\mathrm{rank}_{\mathcal{R}}(GG^\times) \geq \mathrm{rank}_{\mathbf{C}}(DD^\ddagger) \geq \mathrm{rank}_{\mathbf{C}}(D) = \mathrm{rank}_{\mathcal{R}}(G)$$

and $\mathrm{rank}_{\mathcal{R}}(GG^\times) = \mathrm{rank}_{\mathcal{R}}(G)$. Thus, $\mathrm{im}\,(GG^\times) = \mathrm{im}\,(G)$ and multiplication by $GG^\times$ is a projection onto $\mathrm{im}\,G$. Hence $GG^\times G = G$. □

Theorem 3.1 has been obtained originally [23] as a corollary to a generalization of the theorem of Bart-Gohberg-Kaashoek-Van Dooren on minimal factorization of regular rational matrix functions [7]. The connection between the spectrum of $A^\times$ and the zeros of G has been indicated in [27]. Note that both properties (3.2) of a generalized inverse of a matrix have been used in the proof.

The pointwise generalized inverses of G which arise through formula (3.14) are precisely those which are analytic at infinity and have zero defect [23, 24], where the *defect* of a rational matrix function is the difference between the McMillan degree of the function and the sum of multiplicities of all its zeros [17]. The defect of a function $W \in \mathcal{R}^{m\times n}$ can be also characterized as follows. Let $\Omega\mathcal{R}$ and $\Omega_\infty\mathcal{R}$ be the subrings of $\mathcal{R}$ formed by functions analytic in the complex plane and at infinity, respectively, and denote by $\Omega\mathcal{R}^i$ and $\Omega_\infty\mathcal{R}^i$ appropriate direct sums. Let V be a subspace of $\mathcal{R}^n$. A Wedderburn-Forney space $\mathcal{W}(V)$ associated with V is the finite dimensional space over $\mathbf{C}$ [35, 36]

$$\mathcal{W}(V) = \frac{\pi_- V}{V \cap \pi_- \mathcal{R}^n}$$

where π_- is the projection onto $z^{-1}\Omega_\infty\mathcal{R}^n$ along $\Omega\mathcal{R}^n$. A function $G \in \mathcal{R}^{m\times n}$ has zero defect if and only if $\mathcal{W}(\ker G) = (0)$ and $\mathcal{W}(\mathrm{im}\,G) = (0)$.

Let

$$Z(G) = \frac{G^{-1}(\Omega\mathcal{R}^m) + \Omega\mathcal{R}^n}{\ker G + \Omega\mathcal{R}^n}$$

and

$$Z_\infty(G) = \frac{G^{-1}(\Omega_\infty\mathcal{R}^m) + \Omega_\infty\mathcal{R}^n}{\ker G + \Omega_\infty\mathcal{R}^n}$$

be the zero modules of G in the finite plane and at infinity, and let

$$\mathcal{Z}(G) = Z(G) \oplus Z_\infty(G)$$

be the global zero space of G. Let

$$P(G) = \frac{\Omega\mathcal{R}^n}{G^{-1}(\Omega\mathcal{R}^m) \cap \Omega\mathcal{R}^n}$$

and

$$P_\infty(G) = \frac{\Omega_\infty\mathcal{R}^n}{G^{-1}(\Omega_\infty\mathcal{R}^m) \cap \Omega_\infty\mathcal{R}^n}$$

be the pole modules of G in the finite plane and at infinity, and let $\mathcal{P}(G) = P(G) \oplus P_\infty(G)$ be the global pole space of G. Then [35, 36] there exists an exact sequence of vector spaces and vector space homomorphisms

$$0 \to \mathcal{Z}(G) \to \frac{\mathcal{P}(G)}{\mathcal{W}(\ker G)} \to \mathcal{W}(\operatorname{im} G) \to 0. \tag{3.15}$$

Thus, G has zero defect if and only if $\mathcal{Z}(G) \cong \mathcal{P}(G)$.

The assumption essential in Theorem 3.1 is that G have neither a pole nor zero at infinity. The functions G and H in (2.8′) are strictly proper. Also, equation (2.8′) can be of interest in the case when G or H have a pole at infinity. To write down a formula for a generalized pseudoinverse of an arbitrary rational matrix function, we need a more general representation tool. An arbitrary function $W \in \mathcal{R}^{m\times n}$ can be represented in the generalized state space form [34]

$$W(z) = D + C(zE - A)^{-1}B. \tag{3.16}$$

In fact [34], matrices A, B, C and E can be chosen so that $D = 0$.

The function W can be also represented in the form [28, 11, 1]

$$W(z) = C_C(z - A_C)^{-1}B_C + D_\infty + C_\infty(z^{-1} - A_\infty)^{-1}B_\infty \tag{3.17}$$

with A_∞ nilpotent, where the last two terms realize the polynomial part of W. In [15], nonproper rational matrix functions have been represented in the form

$$W(z) = D + (z - \alpha)C(zG - A)^{-1}B, \tag{3.18}$$

where α is a point in the complex plane. The representation (3.18) has been modified in [25] to the form

$$W(z) = D + C\left(\frac{1}{z-\alpha} - \tilde{A}\right)^{-1}\tilde{B}, \tag{3.19}$$

called a *centered realization*. It is written as $(\tilde{A}, \tilde{B}, C, D, \alpha)$.

A realization (A, B, C, D, α) of a rational matrix function W is *minimal* if the dimension of the state space (i.e. the domain of A) equals the McMillan degree of W. For a fixed α, a minimal realization is unique up to similarity. We note that centered realizations obey the usual rules of realization arithmetic. That is, if $(A_i, B_i, C_i, D_i, \alpha)$ is a realization of a function G_i $(i = 1, 2)$ and the number of columns of G_1 equals the number of rows of G_2, then

$$\left(\begin{bmatrix} A_1 & B_1C_2 \\ 0 & A_2 \end{bmatrix}, \begin{bmatrix} B_1D_2 \\ B_2 \end{bmatrix}, \begin{bmatrix} C_1 & D_1C_2 \end{bmatrix}, D_1D_2, \alpha\right)$$

is a realization of the function G_1G_2. Also, if G_1 and G_2 have equal sizes,

$$\left(\begin{bmatrix} A_1 & 0 \\ 0 & A_2 \end{bmatrix}, \begin{bmatrix} B_1 \\ B_2 \end{bmatrix}, \begin{bmatrix} C_1 & C_2 \end{bmatrix}, D_1 + D_2, \alpha\right)$$

is a realization of the function $G_1 + G_2$.

The advantage of representing nonproper rational matrix functions by means of centered realizations is that centered realizations involve a single state space operator rather than the operator pencil. We recall from [25] the connection between the matrices in (3.17) and those in (3.19).

Proposition 3.2 *Suppose* (3.17) *holds with* A_∞ *nilpotent. Pick a point* $\alpha \in \mathbf{C}$ *such that the matrices* $(\alpha - A_{\mathrm{C}})$ *and* $(I - \alpha A_\infty)$ *are invertible, and let*

$$A = \begin{bmatrix} -(\alpha - A_{\mathrm{C}})^{-1} & \\ & (I - \alpha A_\infty)^{-1}A_\infty \end{bmatrix}, \quad B = \begin{bmatrix} -(\alpha - A_{\mathrm{C}})^{-2}B_{\mathrm{C}} \\ (I - \alpha A_\infty)^{-2}B_\infty \end{bmatrix},$$

$$C = \begin{bmatrix} C_{\mathrm{C}} & C_\infty \end{bmatrix},$$

$D = D_\infty + C_{\mathrm{C}}(\alpha - A_{\mathrm{C}})^{-1}B_{\mathrm{C}} + \alpha\mathbf{C}_\infty(I - \alpha A_\infty)^{-1}B_\infty$. *Then*

$$W(z) = D + C\left(\frac{1}{z-\alpha} - A\right)^{-1}B. \tag{3.20}$$

For the sake of completeness, we include the computation which underlies Proposition 3.2. Suppose (3.17) holds, and let $\alpha \in \mathbf{C}$ be such that the matrices $(\alpha - A_{\mathrm{C}})$ and $(I - \alpha A_\infty)$ are invertible. Then [18]

$$W(z) = D_\infty + C_{\mathbf{C}}(\alpha - A_{\mathbf{C}})^{-1}B_{\mathbf{C}} + \alpha C_\infty(I - \alpha A_\infty)^{-1}B_\infty$$

$$+(z-\alpha)\begin{bmatrix} C_{\mathbf{C}} & C_\infty \end{bmatrix}\left(z\begin{bmatrix} I & 0 \\ 0 & A_\infty \end{bmatrix} - \begin{bmatrix} A_{\mathbf{C}} & 0 \\ 0 & I \end{bmatrix}\right)^{-1}$$
$$\begin{bmatrix} -(\alpha - A_{\mathbf{C}})^{-1}B_{\mathbf{C}} \\ -(I-\alpha A_\infty)^{-1}B_\infty \end{bmatrix}$$
$$:= D + (z-\alpha)C(zG - A_\alpha)^{-1}B_\alpha$$

where the matrix $\alpha G - A_\alpha$ is invertible. Hence

$$\begin{aligned} W(z) &= D + C\left(\frac{z}{z-\alpha}G - \frac{1}{z-\alpha}A_\alpha\right)^{-1} B_\alpha \\ &= D + C\left(G + \frac{\alpha}{z-\alpha}G - \frac{1}{z-\alpha}A_\alpha\right)^{-1} B_\alpha \\ &= D + C\left(\frac{1}{z-\alpha}(\alpha G - A_\alpha) + G\right)^{-1} B_\alpha \\ &= D + C\left(\frac{1}{z-\alpha} + (\alpha G - A_\alpha)^{-1}G\right)^{-1} (\alpha G - A_\alpha)^{-1}B_\alpha \\ &:= D + C\left(\frac{1}{z-\alpha} - A\right)^{-1} B. \end{aligned}$$

Proposition 3.2 has the following converse.

Proposition 3.3 *Let (A, B, C, D, α) be a realization of a function $W \in \mathcal{R}^{m\times n}$. Let S be a nonsingular matrix such that*

$$SAS^{-1} = \begin{bmatrix} A_{11} & 0 \\ 0 & A_{22} \end{bmatrix}$$

with the matrix A_{11} invertible and A_{22} nilpotent, and partition

$$SB = \begin{bmatrix} \tilde{B}_1 \\ \tilde{B}_2 \end{bmatrix}, \quad CS^{-1} = \begin{bmatrix} C_1 & C_2 \end{bmatrix}$$

conformably. Then

$$W(z) = D_1 + C_1(z - A_1)^{-1}B_1 + C_2\left(\frac{1}{z} - A_2\right)^{-1} B_2,$$

where

$$\begin{aligned} A_1 &= \alpha + A_{11}^{-1}, & A_2 &= A_{22}(I + \alpha A_{22})^{-1}, \\ B_1 &= -A_{11}^{-2}\tilde{B}_1, & B_2 &= (I + \alpha A_{22})^{-2}\tilde{B}_2, \end{aligned}$$

and $D_1 = D - C_1 A_{11}^{-1}\tilde{B}_1 - \alpha C_2(I + \alpha A_{22})^{-1}\tilde{B}_2.$

In [23], the formulas for a generalized pseudoinverse of an arbitrary rational matrix function has been given in terms the realization (3.18). The formulas take the following form [24, 10] for a realization (3.20).

Theorem 3.4 *Let $W \in \mathcal{R}^{m\times n}$, pick a point α which is neither a pole nor zero of W, and compute a centered realization (A, B, C, D, α) of W. Then*

$$W^{\times}(z) = D^{\ddagger} - D^{\ddagger}C\left(\frac{1}{z-\alpha} - A^{\times}\right)^{-1} BD^{\ddagger},$$

with $A^{\times} = A - BD^{\ddagger}C$, is a generalized pseudoinverse of W whenever $D^{\ddagger}$ is a generalized inverse of D.

4. Parametrization of all disturbance decoupling controllers. Theorem 3.4 and formula (3.11) provide the general solution of equation (2.9′). To characterize the proper solutions, we will need certain elements of valuation theory. The valuation-theoretic tools have been frequently used in systems theory [13, 33, 17]. We recall now the definition of orthogonality in $\mathcal{R}^n$ adopted in [2, 3].

Pick a point $\lambda \in \mathbf{C}_\infty$. Let $|0|_\lambda = 0$, and if r is a nonzero element of $\mathcal{R}$, let

$$|r|_\lambda = e^{\eta} \tag{4.1}$$

where η is the integer such that $(z-\alpha)^{\eta} r(z)$ (or $z^{-\eta}r(z)$ if $\lambda = \infty$) is analytic, and does not vanish, at λ. The function $|\cdot|_\lambda$ determines the non-Archimedean valuation of $\mathcal{R}$. If $x = (x_1, x_2, \ldots, x_n) \in \mathcal{R}^n$, let

$$||x||_\lambda = \max\{|x_1|_\lambda, |x_2|_\lambda, \ldots, |x_n|_\lambda\}. \tag{4.2}$$

Then $(\mathcal{R}^n, ||\cdot||_\lambda)$ is a non-Archimedean normed space over the real valued field $(\mathcal{R}, |\cdot|_\lambda)$. The space $(\mathcal{R}^n, ||\cdot||_\lambda)$ is not complete, but the concept of orthogonality [20] still applies. If U and V are subspaces of $\mathcal{R}^n$,

$$||u+v||_\lambda \le ||u||_\lambda + ||v||_\lambda \tag{4.3}$$

for all $u \in U$ and $v \in V$. Subspaces U and V are orthogonal with respect to $||\cdot||_\lambda$ if (4.3) is always an equality. If this happens, we will say then that U and V are *orthogonal at* λ. Subspaces U and V of $\mathcal{R}^n$ are orthogonal on a set $\sigma \in \mathbf{C}_\infty$ if they are orthogonal at every point of σ.

Orthogonality of subspaces U and V of $\mathcal{R}^n$ at λ can be characterized in terms of linear algebra as follows. If Ω is a subspace of $\mathcal{R}^n$, let

$$\Omega(\lambda) = \{\phi(\lambda) : \ \phi \in \Omega \text{ is analytic at } \lambda\}. \tag{4.4}$$

Then $\Omega(\lambda)$ is a subspace of $\mathbf{C}^n$, and

$$\dim_{\mathbf{C}}\Omega(\lambda) = \dim_{\mathcal{R}}\Omega. \tag{4.5}$$

The space $\Omega(\lambda)$ has been called the *value* of Ω at λ. It can be alternatively characterized as the linear span of the leading coefficients in the Laurent expansions at λ of the functions in Ω. The subspaces U and V of $\mathcal{R}^n$ are orthogonal at λ [2, 3] if

$$U(\lambda) \cap V(\lambda) = (0). \tag{4.6}$$

Let $G^\times$ be a pointwise generalized inverse of a function $G \in \mathcal{R}^{m\times n}$. Then

$$WW^\times WW^\times = WW^\times$$

and the map $\pi : \mathcal{R}^m \to \mathcal{R}^n$ defined by

$$\pi\phi = WW^\times\phi \tag{4.7}$$

is a projection onto the subspace U spanned by the columns of $WW^\times$ along the subspace spanned by the columns of $(I - WW^\times)$. Since $\operatorname{rank}_\mathcal{R} W = \operatorname{rank}_\mathcal{R}(WW^\times)$, V is spanned by the columns of W. Similarly, postmultiplication by $WW^\times$ is a projection of $\mathcal{R}^{1\times n}$ onto the row span of $W^\times$ along the subspace spanned by the rows of $(I - WW^\times)$.

Pick complementary subspaces U and V of $\mathcal{R}^n$, and let π be the projection onto U along V. The projection π is said to be *orthogonal* at λ [2] if U and V are orthogonal at λ. By Proposition 2.1 in [26] (see also Proposition 4.3 in [2]), the following statements are equivalent:

(i) the projection π is orthogonal at λ,
(ii) $\pi(\phi)$ is analytic at λ whenever ϕ is analytic at λ.

Orthogonal projections corresponding to multiplication by $WW^\times$ can be also characterized as follows [26]. If $W \in \mathcal{R}^{m\times n}$, let

$$W^{o\ell} = \{\phi \in \mathcal{R}^{1\times m} :\ \phi W = 0\} \tag{4.8}$$

and let

$$W^{or} = \{\psi \in \mathcal{R}^{n\times 1} :\ W\psi = 0\}. \tag{4.9}$$

Let EGF and $\overline{E}\,\overline{G}\,\overline{F}$ be the Smith's normal forms [1] at λ of W and $W^\times$, respectively. Suppose the normal rank of W (and $W^\times$) equals k. Then

$$EGF = \begin{bmatrix} E_1 & E_2 \end{bmatrix} \begin{bmatrix} G_1 & 0 \\ 0 & 0 \end{bmatrix} \begin{bmatrix} F_1 \\ * \end{bmatrix} = E_1G_1F_1$$

where G_1 has the size $k \times k$. Similarly, $\overline{E}\,\overline{G}\,\overline{F} = \overline{E}_1\,\overline{G}_1\,\overline{F}_1$. From the first equality in (3.10) it follows that

$$G_1(z)F_1(z)\overline{E}_1(z)\overline{G}_1(z)\overline{F}_1(z)E_1(z) = I,$$

and so $G_1F_1\overline{E}_1\,\overline{G}_1$ is the pseudoinverse of $\overline{F}_1E_1$. Now E_1 and $\overline{F}_1$ are analytic at λ, the column span of E_1 is equal to the column span of W, and the row span of $\overline{F}_1$ is equal to the row span of $W^\times$. Consequently, the matrix

$\overline{F}_1(\lambda)E_1(\lambda)$ is nonsingular if and only if $W^{\times or}$ and the column span of W are orthogonal at λ. Therefore, the function $WW^\times = E_1G_1F_1\overline{E}_1\overline{G}_1\overline{F}_1$ is analytic at λ if and only if $W^{\times or}$ and the column span of W are orthogonal at λ. Similarly, the function $W^\times W$ is analytic at λ if and only if the row span of W is orthogonal to $W^{\times o\ell}$ at λ. We summarize the results of the preceding discussion.

Proposition 4.1 *Let $W^\times \in \mathcal{R}^{n\times m}$ be a generalized pseudoinverse of a function $W \in \mathcal{R}^{m\times n}$, let π_L and π_R be the projections*

$$\pi_L(\phi) = W^\times W\phi \qquad \text{and} \qquad \pi_R(\psi) = \psi WW^\times,$$

and let $\lambda \in \mathbf{C}_\infty$. Then

(i) *π_L is orthogonal at λ if and only if the row span of W is orthogonal to $W^{\times o\ell}$ at λ,*

(ii) *π_R is orthogonal at λ if and only if the column span of W is orthogonal to $W^{\times or}$ at λ.*

The generalized pseudoinverse (3.14) of the function $G \in \mathcal{R}^{m\times n}$ has zero defect; $G^{\times or}(\lambda)$ and $G^{\times o\ell}(\lambda)$ coincide with the right and left kernels of $D^\ddagger$ for all $\lambda \in \mathbf{C}_\infty$. Consequently, we can choose a generalized inverse $D^\ddagger$ of D so that the projection determined by left multiplication by $G^\times G$ and/or right multiplication by $GG^\times$ is orthogonal at a given point $\lambda \in \mathbf{C}_\infty$.

We can give now a necessary and sufficient condition for existence of proper solutions of equation (2.8′). Pick a generalized pseudoinverse $G^\times$ of G such that $G^{\times o\ell}$ is orthogonal to the row span of G at infinity, and a generalized pseudoinverse $H^\times$ of H such that $H^{\times or}$ is orthogonal to the column span of H at infinity. Equation (2.8′) is solvable if and only if $GG^\times RH^\times H = R$. If $G^\times RH^\times$ is proper, then obviously equation (2.8′) has a proper solution. Suppose equation (2.8′) has a proper solution Z. Then $G^\times RH^\times$ and Z are two solutions of (2.8′), and so there exists a rational matrix function Y such that

$$Z = G^\times RH^\times + Y - G^\times GYHH^\times. \tag{4.10}$$

Hence

$$G^\times GZHH^\times = G^\times RH^\times, \tag{4.11}$$

and, by Proposition 4.1, the function $G^\times RH^\times$ is proper. Thus, a solvable equation (2.8′) has a proper solution if and only if $G^\times RH^\times$ is proper.

Al proper solutions of equation (2.8′) can be parametrized as follows. Let $G^\times$ and $H^\times$ be as above, and suppose $G^\times RH^\times$ is a proper solution of equation (2.8′). If Y is a proper rational matrix function of appropriate size, then

$$X = G^\times RH^\times + Y - G^\times GYHH^\times \tag{4.12}$$

is a proper solution of equation (2.8′).

Suppose Z is a proper solution of equation (2.8′). Then the function $G^\times RH^\times$ is proper and

$$Z = G^\times RH^\times + \tilde{Y} - G^\times G\tilde{Y}HH^\times \tag{4.13}$$

for some rational matrix function $\tilde{Y}$. Let

$$Y = \tilde{Y} - G^\times G\tilde{Y}HH^\times. \tag{4.14}$$

By Proposition 4.1, $G^\times G$ and, in view of (4.13), the functions

$$(I - G^\times G)\tilde{Y} \qquad \text{and} \qquad \tilde{Y}(I - HH^\times) \tag{4.15}$$

are proper. Hence

$$Y = (I - G^\times G)\tilde{Y} + (G^\times G)\tilde{Y}(I - HH^\times). \tag{4.16}$$

is proper, and

$$\begin{aligned} Z &= G^\times RH^\times + Y \\ &= G^\times RH^\times + Y - G^\times GYHH^\times. \end{aligned}$$

Thus, (4.12) with Y proper is the general proper solution of equation (2.8′).

Theorem 4.2 *Let G, H, and R be rational matrix functions. Pick a point $\alpha \in \mathbf{C}$ which is neither a pole nor a zero of G or H, and compute realizations $(A_G, B_G, C_G, D_G, \alpha)$ and $(A_H, B_H, C_H, D_H, \alpha)$ of G and H. Compute generalized inverses $D_G^\ddagger$ and $D_H^\ddagger$ of D_G and D_H such that the row span of $D_H^\ddagger$ intersects trivially with $H^{o\ell}(\infty)$ and the column span of $D_G^\ddagger$ intersects trivially with $G^{or}(\infty)$. Let*

$$G^\times(z) = D_G^\ddagger - D_G^\ddagger C_G \left(\frac{1}{z-\alpha} - A_G^\times\right)^{-1} B_G D_G^\ddagger,$$

and let

$$H^\times(z) = D_H^\ddagger - D_H^\ddagger C_H \left(\frac{1}{z-\alpha} - A_H^\times\right)^{-1} B_H D_H^\ddagger$$

where $A_G^\times = A_G - B_G D_G^\ddagger C_G$ and $A_H^\times = A_H - B_H D_H^\ddagger C_H$. Then R admits a factorization $R = GXH$ with the factor X causal if and only if $GG^\times RH^\times H = R$ and the function $G^\times RH^\times$ is proper. Moreover, if the condition is satisfied, X is a causal factor if and only if

$$X = G^\times RH^\times + Y - G^\times GYHH^\times$$

for some proper rational matrix function Y.

Theorem 4.2 yields the following parametrization of disturbance decoupling controllers for the plant (2.1).

Theorem 4.3 *Pick a point* $\alpha \in \mathbf{C}\backslash\sigma(A)$ *which is not a zero of* $T_{y\leftarrow u_c}(s) = C(s-A)^{-1}B$ *or* $T_{y_m\leftarrow u_d}(s) = C_m(s-A)^{-1}B_d$. *Let* $A_\alpha = (A-\alpha)^{-1}$, *and let*

$$\begin{array}{llllll} B_G & = -(\alpha - A)^{-2}B, & C_G & = C, & D_G & = C(\alpha - A)^{-1}B \\ B_H & = -(\alpha - A)^{-2}B_d, & C_H & = C_m, & D_H & = C(\alpha - A)^{-1}B_d \\ B_R & = -(\alpha - A)^{-2}B, & C_R & = C, & D_R & = C(\alpha - A)^{-1}B_d. \end{array}$$

Choose a generalized inverses $D_G^\ddagger$ *and* $D_H^\ddagger$ *of* D_G *and* D_H *such that the column span of* $D_G^\ddagger$ *intersects trivially with* $T^{\mathrm{or}}_{y\leftarrow u_c}(\infty)$ *and the row span of* $D_H^\ddagger$ *intersects trivially with* $T^{o\ell}_{y_m\leftarrow u_d}(\infty)$. *Then there exists a disturbance decoupling compensator if and only if*

$$GG^\times RH^\times H = R$$

and the function $G^\times RH^\times$ *is proper, where*

$$\begin{aligned} G(z) &= D_G + C_G\left(\frac{1}{z-\alpha} - A_\alpha\right)^{-1} B_G, \\ G^\times(z) &= D_G^\ddagger - D_G^\ddagger C_G\left(\frac{1}{z-\alpha} - A_\alpha + B_G D_G^\ddagger C_G\right)^{-1} B_G D_G^\ddagger, \\ H(z) &= D_H + C_H\left(\frac{1}{z-\alpha} - A_\alpha\right)^{-1} B_H, \\ H^\times(z) &= D_H^\ddagger - D_H^\ddagger C_H\left(\frac{1}{z-\alpha} - A_\alpha + B_H D_H^\ddagger C_H\right)^{-1} B_H D_H^\ddagger, \end{aligned}$$

and

$$R(z) = D_R + C_R\left(\frac{1}{z-\alpha} - A_\alpha\right)^{-1} B_R.$$

Moreover, suppose there exists a disturbance decoupling compensator and $T_{y_m\leftarrow u_c}(s) = C_m(s-A)^{-1}B$.

Then T_c is the transfer function of a disturbance decoupling compensator if and only if

$$T_c = (G^\times RH^\times + Y - G^\times GYHH^\times)(T_{y_m\leftarrow u_c}(G^\times RH^\times + Y - G^\times GYHH^\times) - I)^{-1}$$

for some proper rational matrix function Y.

REFERENCES

[1] J. A. Ball, I. Gohberg and L. Rodman, *Interpolation of Rational Matrix Functions*, OT 45, Birkhäuser Verlag, Basel Boston Berlin, 1990.

[2] J.A. Ball and M. Rakowski, Zero-Pole Structure of Nonregular Rational Matrix Functions, in *Extension and Interpolation of Linear Operators and Matrix Functions*, OT 47, Birkhäuser Verlag, Basel Boston Berlin, pp. 137-193, 1990.

[3] ———, Null-Pole Subspaces of Rectangular Rational Matrix Functions, *Lin. Alg. Appl.*, 159 (1991), pp. 81-120.

[4] ———, Interpolation by rational matrix functions and stability of feedback systems: The 2-block case, *J. Math. Systems, Estimation, and Contr.*, to appear.

[5] ———, Interpolation by rational matrix functions and stability of feedback systems: the 4-block case, in *Proceedings of Workshop on Operator Theory and Complex Analysis*, Hokkaido Univ., 1991, OT 59, Birkhäuser Verlag, Basel, pp. 96–142, 1992.

[6] H. Bart, I. Gohberg and M. A. Kaashoek, *Minimal Factorization of Matrix and Operator Functions*, OT 1, Birkhäuser Verlag, Basel-Boston-Stuttgart, 1979.

[7] H. Bart, I. Gohberg, M. A. Kaashoek and P. Van Dooren, Factorization of Transfer Functions, *SIAM J. Contr. and Opt.*, 18 (1980), pp. 675-696.

[8] A. Ben-Israel and T. N. E. Greville, *Generalized Inverses: Theory and Applications*, John Wiley & Sons, New York London Sydney Toronto, 1974.

[9] S. L. Campbell and C. D. Meyer, Jr., *Generalized Inverses of Linear Transformations*, Dover Press, 1991.

[10] S. L. Campbell and M. Rakowski, Explicit Formulae for Completions of Linear Time-Varying Singular Systems of Differential Equations, *Circ., Syst., and Sig. Proc.*, to appear.

[11] N. Cohen, On Spectral Analysis and Factorization of Rational Matrix Functions, Ph.D. Dissertation, Weizmann Institute of Science, Rehovot, Israel, August 1984.

[12] G. Conte and A. M. Perdon, On the Causal Factorization Problem, in *IEEE Trans. Aut. Contr.*, AC-30 (1985), pp. 811-813.

[13] G. D. Forney, Jr., Minimal bases of rational vector spaces, with applications to multivariable linear systems, *SIAM J. Contr.*, 13 (1975), pp. 493-520.

[14] B. Francis, *A course in* H_∞ *Control Theory*, Springer Verlag, Berlin Heidelberg New York, 1987.

[15] I. Gohberg and M. A. Kaashoek, Regular Rational Matrix Functions with Prescribed Pole and Zero Structure, in *Topics in Interpolation Theory of Rational Matrix Functions*, OT 33, Birkhäuser Verlag, Basel-Boston-Berlin, pp. 109-122, 1988.

[16] J. Hammer and M. Heyman, Causal factorization and linear feedback, *SIAM J. Contr. Opt.*, 19 (1981), pp. 445-468.

[17] T. Kailath, *Linear Systems*, Prentice-Hall, Englewood Cliffs, N. J., 1980.

[18] J. Kim-Kang, Interpolation by Rational Matrix Functions with Minimal McMillan degree, Ph.D. Dissertation, Virginia Tech, Blacksburg, Virginia, April 1990.

[19] S. Kung and T. Kailath, Some Notes On Valuation Theory In Linear Systems, *Proc. of the 1978 IEEE Conf. on Decision and Control, San Diego, CA*, pp. 515-517, 1978.

[20] A. F. Monna, *Analyse non-archimédienne*, Springer-Verlag, Berlin Heidelberg New York, 1970.

[21] D. Y. Ohm, S. P. Bhattacharyya and J. W. Howze, Transfer Function Conditions for (C', A, B)-pairs, *IEEE Trans. Automat. Contr.*, AC-29 (1984), pp. 172-174.

[22] R. Penrose, A generalized inverse for matrices, *Proc. Cambridge Philos. Soc.*, 51 (1955), pp. 406-413.

[23] M. Rakowski, Generalized Pseudoinverses of Matrix Valued Functions, *Int. Eq. and Op. Th.*, 14 (1991), pp. 564-585.

[24] ———, Zero-Defect Generalized Pseudoinverses of Rational Matrix Functions, in *Recent Advances in Mathematical Theory of Systems, Control, Networks and Signal Processing I*, Mita Press, Tokyo, pp. 197-202, 1992.

[25] ———, Minimal Factorization of Rational Matrix Functions, *IEEE Trans. Circ. and Syst.-I: Fund. Th. Appl.*, CAS-39 (1992), pp. 440-445.

[26] ———, State-Space Solution of the Causal Factorization Problem, *IEEE Trans. Aut. Contr.*, to appear.

[27] ———, Spectral Factorization of Rectangular Rational Matrix Functions with Application to Discrete Wiener-Hopf Equations, *J. Func. Anal.*, vol. 110 (1992), pp. 410–433.

[28] H. H. Rosenbrock, *State-space and Multivariable Theory*, John Wiley & Sons, Inc., New York, 1970.

[29] J. M. Schumacher, Compensator Synthesis Using (C, A, B)-Pairs, *IEEE Trans. Automat. Contr.*, AC-25 (1980), pp. 1133-1138.

[30] ———, Regulator Synthesis Using (C, A, B)-Pairs, *IEEE Trans. Automat. Contr.*, AC-27 (1982), pp. 1211-1221.

[31] L. M. Silverman, Inversion of Multivariable Linear Systems, *IEEE Trans. Auto. Contr.*, AC-14 (1969), pp. 270-276.

[32] A. E. Taylor and D. C. Lay, *Introduction to Functional Analysis*, Second Edition, John Wiley & Sons, New York Chichester Brisbane Toronto, 1980.

[33] G. Verghese and T. Kailath, Rational Matrix Structure, in *Proc. 18 IEEE Conf. on Decision and Contr., Fort Lauderdale, FL, Dec. 12-14, 1979*, pp. 1008-1012.

[34] G. Verghese, P. Van Dooren and T. Kailath, Properties of the system matrix of a generalized state-space system, *Int. J. of Contr.*, 30 (1979), pp. 235-243.

[35] B. F. Wyman, M. K. Sain, G. Conte and A. M. Perdon, On the Zeros and Poles of a Transfer Function, *Lin. Alg. and its Appl.*, 122/123/124 (1989), pp. 123-144.

[36] ———, B. F. Wyman, M. K. Sain, G. Conte and A. M. Perdon, Poles and Zeros of Matrices of Rational Functions, *Lin. Alg. and its Appl.*, 157 (1991), pp. 113-139.

SOME NUMERICAL CHALLENGES IN CONTROL THEORY

PAUL VAN DOOREN*

Abstract. We discuss a number of novel issues in the interdisciplinary area of numerical linear algebra and control theory. Although we do not claim to be exhaustive we give a number of problems which we believe will play an important role in the near future. These are : sparse matrices, structured matrices, novel matrix decompositions and numerical shortcuts. Each of those is presented in relation to a particular (class of) control problems. These are respectively : large scale control systems, polynomial system models, control of periodic systems, and normalized coprime factorizations in robust control.

Key words. Numerical algorithms, linear algebra, sparse matrices, polynomial systems, periodic systems, robust control

AMS(MOS) subject classifications. 93b55

1. Introduction. During the last few decades linear algebra has played an important role in advances being made in the area of systems and control [32]. The most profound impact has been in the computational and implementational aspects, where numerical linear algebraic algorithms have strongly influenced the ways in which problems are being solved. This paper discusses a number of novel numerical linear algebra issues in this interdisciplinary area. Although we do not claim to be exhaustive we give a number of upcoming issues which we believe will play an important role in the near future. These numerical linear algebra issues are :

- **Sparse matrices**. Large plants in control typically arise from discretizations of continuum problems (such as those that appear in mechanics or chemistry). The models obtained from that are then automatically sparse, such as finite element methods used for mechanical problems. It is also commonly observed that large scale problems in other application areas typically involve matrices with some kind of sparsity or structure. A recent report on large matrix problems [9] shows that indeed there are *no large general dense matrix problems* to be tackled in most application areas.
- **Structured matrices**. A compact way of representing the transfer function of a system is by using polynomial system models. The number of parameters in these models is often much lower than for their equivalent state space model. Numerical linear algebra problems occurring there typically involve matrix problems with some kind of structure (Hankel, Toeplitz, companion, ...) and the solution of the underlying control problem usually boils down to some

* University of Illinois at Urbana-Champaign, Coordinated Science Laboratory, 1308 W. Main Str., Urbana, IL 61801, Phone: (217)3330656, Fax: (217)2441653, E-mail: vdooren@uicsl.csl.uiuc.edu.

matrix decomposition of such a structured matrix.

- **New decompositions**. The most reliable numerical linear algebra methods proposed for particular control problems are related to particular eigenvalue and singular value decompositions of "special" matrices, such as the Schur decomposition of a Hamiltonian matrix for solving Riccati equations. Several new decomposition have been proposed recently in numerical linear algebra and their application to control theoretic problems still has to be explored.
- **Numerical shortcuts**. A typical approach for solving matrix problems for which there is no "direct" algorithm, is to break it down into a sequence of "intermediate" problems for which one can apply known techniques. A typical early example of this was the construction of the normal equations for solving least squares problems. This is nowadays replaced by the "shortcut" given by the QR decomposition, which is known to be more reliable in general. Similar "detours" are still present in the solution of certain control problems and ought to be avoided if possible.

In the sections below we elaborate on these topics and give for each a particular control theoretic example where it applies. These control examples illustrate well the issues being raised.

2. Sparse matrices and large scale systems. Large plants in control typically arise from discretizations of continuum problems such as finite element methods used for mechanical problems. The models used are then almost always represented in state space (or generalized state space) form. The plant matrices $\{A, B, C, D\}$ typically have a special structure such as sparsity (due to modeling techniques using e.g. finite element methods) or diagonal dominance (inherited from the physical properties of the system being modeled [22], [27]).

For large scale systems, the design of the controller is often faced with the practical constraint that it has to be of relatively low order. The main reasons for this are that the controller has to run in real time at a specific clock rate or has to be implemented on existing hardware of limited capacity (a typical example of this is the laser beam tracking device of the compact disc player which has to be implemented on an existing digital signal processor with relatively small memory and processing speed [28]). But even when no such constraint is imposed, model reduction is a useful approach for yielding approximate solutions to particular control problems. Instead of solving the control problem for the large scale system one solves the corresponding problem for its lower order approximation and then tries to derive from that an *approximate solution* for the original problem.

A justification for this follows from the following observation. A large class of model reduction methods of a system given in state space form $\{A, B, C, D\}$ can be interpreted as performing a similarity transformation T yielding $\{A^{(t)}, B^{(t)}, C^{(t)}, D\} \doteq \{T^{-1}AT, T^{-1}B, CT, D\}$, and then extracting from that the leading diagonal system $\{\hat{A}, \hat{B}, \hat{C}, D\}$, i.e.

$$\left[\begin{array}{c|c} T^{-1}AT & T^{-1}B \\ \hline CT & D \end{array}\right] \doteq \left[\begin{array}{cc|c} A_{1,1}^{(t)} & A_{1,2}^{(t)} & B_1^{(t)} \\ A_{2,1}^{(t)} & A_{2,2}^{(t)} & B_2^{(t)} \\ \hline C_1^{(t)} & C_2^{(t)} & D \end{array}\right] \tag{2.1}$$

$$\left[\begin{array}{c|c} \hat{A} & \hat{B} \\ \hline \hat{C} & \hat{D} \end{array}\right] \doteq \left[\begin{array}{c|c} A_{1,1}^{(t)} & B_1^{(t)} \\ \hline C_1^{(t)} & D \end{array}\right].$$

When writing $T \doteq [X_1 \mid X_2]$, $T^{-1} \doteq [Y_1^t \mid Y_2^t]^t$, then $\Pi \doteq X_1Y_1$ is a projector on X_1 along Y_1, and the subsystem $\{\hat{A}, \hat{B}, \hat{C}, D\}$ is in fact the original system restricted to that projector : $\{\Pi A\Pi, \Pi B, C\Pi, D\}$. Such model approximations thus only differ in the choice of projector. Each choice typically tries to achieve some kind of decomposition. Modal approximation e.g. selects "dominant" frequencies in A and performs thus a block diagonal decomposition (1), i.e. with $A_{1,2}^{(t)}$ and $A_{2,1}^{(t)}$ being 0 and $A_{1,1}^{(t)}$ containing the dominant eigenfrequencies. When large scale systems $\{A, B, C, D\}$ are involved one can only afford approximate decompositions, i.e. with $A_{1,2}^{(t)}$ and $A_{2,1}^{(t)}$ small compared to $A^{(t)}$. Iterative techniques for large systems are indeed only applied a limited number of steps for reasons of complexity. Techniques for nearly decomposed systems [27], [22] can then be used to derive bounds on the approximation criterion or an iterative refinement to improve the decomposition.

The above ideas thus suggest the following approach for solving particular control problems for large scale systems :

- Choose a projector Π yielding a lower order system $\{\Pi A\Pi, \Pi B, C\Pi, D\}$
- Solve the given problem for this lower order system using dense matrix techniques
- "Lift" back the solution to the original coordinate system.

This sounds of course very simple but the crux is to find the projector Π that achieves two important goals. It should be *easy to construct*, i.e. one should be able to exploit sparsity of the model $\{A, B, C, D\}$ for the construction of Π. And the lifted solution should provide a *good approximation* for the true solution of the control problem, i.e. certain "performance bounds" ought to be satisfied. For the first goal there exist a number of iterative techniques such as Krylov type schemes (Lanczos, Arnoldi, GMRES, QMR [25], [11]). The second goal is of course problem dependent and is probably the most challenging one. One may have to assume certain system properties here (such as diagonal dominance) in order to derive sufficiently powerful results. Also criteria have to be selected to obtain an appropriate system performance and how far iteration techniques have to be applied in order to satisfy corresponding bounds. The complexity of this approach typically ought to be an order of magnitude less (in terms of the system order n) than the complexity of dense matrix techniques applied to the original problem.

We give here an example of a projection construction based on the Quasi Minimal Residual (or QMR) algorithm [11] [26]. Consider a $n \times n$ matrix A and two starting vectors $v_1 = b$ and $w_1 = c$. The two sided Lanczos algorithm then essentially constructs column by column matrices $V_m = [v_1 ... v_m]$ and $W_m = [w_1 ... w_m]$ satisfying

$$\begin{aligned} AV_m &= V_m T_m + \delta_{m+1} v_{m+1} e_m^T \\ A^T W_m &= W_m T_m^T + \beta_{m+1} w_{m+1} e_m^T \\ V_m^T W_m &= I_m, \end{aligned} \tag{2.2}$$

where T_m is a tridiagonal matrix of dimension $m \times m$ and V_m and W_m are $n \times m$ dense matrices. The construction of the matrices V_m, W_m and T_m only involves matrix multiplies of the type Av and $A^T w$ and is thus "cheap" when A is sparse and $m << n$. This algorithm may break down (e.g. when the bi-orthonogality condition $V_m^T W_m = I_m$ can not be satisfied) but a modification called the QMR algorithm [11] was recently proposed that avoids these pitfalls. An interesting result of these "bases" V_m and W_m is that they provide good approximate solutions for the underlying Lyapunov equations

$$\begin{aligned} AP_c + P_c A^T &= bb^T \\ A^T P_o + P_o A &= c^T c. \end{aligned} \tag{2.3}$$

Based on these approximate solutions for P_o and P_c, one derives a reduced order model that in fact corresponds to a projector $\Pi \doteq V_m W_m^T$. Approximation bounds are linked to approximations of the matrix exponentials $c^T e^{At}$, $e^{At} b$ and $c^T e^{At} b$ [25]. Extensions to the block Lanczos case and various bounds for the resulting lower order approximation are still under investigation.

3. Structured matrices and polynomial system models. Polynomial system models have been proposed for most control problems as an interesting alternative to state space techniques [2], [23]. The key advantage of this approach is the reduced complexity of the algorithms for solving particular problems. In turn, the main reason for that is the reduced number of parameters needed to represent a system of the same order. As an example, a n-th order SISO (single input single output) system requires $O(n^2)$ parameters when represented by an arbitrary state space model $\{A, B, C, D\}$, while it requires only $O(n)$ parameters when represented by a polynomial model $\{p(z), q(z)\}$.

Consider then the problem of checking minimality of the given realization. In both cases this amounts to a rank test. For the state space realization one has to check the rank of the controllability and observability matrices, which requires $O(n^3)$ operations. For the polynomial model one has to check coprimeness of $p(z)$ and $q(z)$, which amounts to checking the rank of the Sylvester matrix [2], [12] :

$$S_{2n} \doteq \begin{bmatrix} p_0 & p_1 & \dots & p_n & 0 & \dots & 0 \\ 0 & p_0 & p_1 & \dots & p_n & \ddots & \vdots \\ \vdots & \ddots & \ddots & \ddots & & \ddots & 0 \\ 0 & \dots & 0 & p_1 & p_2 & \dots & p_n \\ 0 & \dots & 0 & q_1 & q_2 & \dots & q_n \\ \vdots & \cdot^{\cdot^{\cdot}} & \cdot^{\cdot^{\cdot}} & \cdot^{\cdot^{\cdot}} & & \cdot^{\cdot^{\cdot}} & 0 \\ 0 & q_0 & q_1 & \dots & q_n & & \vdots \\ q_0 & q_1 & \dots & q_n & 0 & \dots & 0 \end{bmatrix} \tag{3.1}$$

Although this is a $2n \times 2n$ matrix its rank can be computed by Euclid's algorithm and this requires only $O(n^2)$ operations !

This difference of an order n in operation count is typical in the SISO case and thus speaks strongly in favor of polynomial models. Yet, the algorithms for polynomial models are typically known to be numerically unstable (see e.g. [29]). Indeed, Euclid's algorithm can be shown to be equivalent to Gaussian elimination *without pivoting* on the Sylvester matrix, and this is known to be numerically unreliable. Moreover Euclid's algorithm is also at the basis of most polynomial matrix decompositions such as the Smith form or Hermite form of a polynomial matrix. The polynomial approach thus seems to sacrifice accuracy for speed.

We strongly believe that iterative refinement could be a way to overcome the above drawback of the polynomial approach. It is known indeed that a few steps of iterative refinement can often turn an unstable method into a stable one [41] and this ought to be tried out on the matrix agorithm for polynomial models. If the number of refinement steps is still significantly lower than n then one still has an efficient algorithm compared to the corresponding state space algorithms. We now illustrate this with two examples. The first one is the computation of poles of a system. For a state space system this boils down to computing the eigenvalues of the matrix A, and it requires $O(n^3)$ operations using the (numerically stable) QR algorithm. For a polynomial model this amounts to computing the zeros of $q(z)$, but this can be done in $O(n^2)$ operations using a stable algorithm [21] ! The basic iterative scheme underlying this is the Newton-Raphson algorithm which is a kind of iterative refinement.

The second example is a more important one in control applications and is known as *spectral factorization.* It is the corner stone in problems of optimal control and robust control [8], [38]. For a state space model one typically solves this via a Riccati equation which is again an eigenvalue problem (with particular symmetry) and requires $O(n^3)$ operations

[24], [30]. For a (discrete time) SISO polynomial model one can state the problem as follows. Given the finite power series

$$\Phi(z) = \phi_n z^{-n} + \cdots + \phi_1 z^{-1} + \phi_0 + \phi_1 z + \cdots + \phi_n z^n \tag{3.2}$$

find a *stable* polynomial $p(z)$ (i.e. zeros inside $\|z\| < 1$) such that :

$$\Phi(z) = p(z^{-1})\, p(z). \tag{3.3}$$

It is known that this always has a solution if $\Phi(e^{j\omega}) \geq 0$ for all ω.

Vostry's [39] proposed an algorithm for this which starts with some stable polynomial $x^{(0)}(z)$ and then iterates with :

$$p^{(i)}(z^{-1})x^{(i)} + p^{(i)}(z).x^{(i)}(z^{-1}) = 2\Phi(z)$$

$$p^{(i+1)}(z) = \frac{1}{2}[p^{(i)}(z) + x^{(i)}(z)] \tag{3.4}$$

One iteration step requires the solution of a system of equations with a matrix close to the Sylvester matrix and again this can be done in $O(n^2)$ operations. The stability of a single step is *not* guaranteed, but since it is a method of iterative refinement using the original data $\Phi(z)$ for computing the correction, the overall stability of Vostry's method is quite good ! The convergence of $p^{(i)}(z)$ to $p(z)$ is quadratic when no roots of $p(z)$ are close to the unit circle and the overall process is thus also $O(n^2)$. It is in fact a Newton correction scheme applied to equation (3).

4. Novel decompositions and periodic Riccati equations. The most reliable numerical linear algebra methods proposed for particular control problems are related to particular eigenvalue and singular value decompositions of "special" matrices, such as the Schur decomposition of a Hamiltonian matrix for solving Riccati equations [24], [30]. Several new decompositions have been proposed recently in numerical linear algebra and their application to control theoretic problems still has to be fully exploited. As an example the generalized Schur form has been shown to have several applications in geometric systems theory [29], and this decomposition is an extension of the standard Schur form to singular pencils of matrices.

In [3] a new decomposition for a sequence of matrices A_i, B_i, $i = 1, \ldots, K$ is proposed. Consider the set of (homogenous) difference equations

$$B_i \cdot x_{i+1} = A_i \cdot x_i, \quad i = 1, \ldots \tag{4.1}$$

with *periodic coefficients* $A_i = A_{i+K}$, $B_i = B_{i+K}$. For period $K = 1$ one has the constant coefficient case $A_i = A$, $B_i = B$ and it is well-known that the generalized eigenvalues of the pair A, B yields important information of the system (6). When $K > 1$ one can derive from (6) a set of K time invariant systems which describe completely the behavior of (6). For simplicity we assume all B_i to be invertible. Then define the matrices $S_i = B_i^{-1}A_i$ yielding the explicit system :

$$x_{i+1} = B_i^{-1}A_i \cdot x_i = S_i \cdot x_i, \quad i = 1, \ldots \tag{4.2}$$

and using $S^{(k)} = S_{k+K-1} \cdot \ldots S_{k+1} \cdot S_k$, $k = 1, ..., K$ the set of K *time invariant* but *subsampled systems* :

$$\begin{aligned} x_{1+(i+1)K} &= S^{(1)} \cdot x_{1+iK}, \quad i = 1, 2, \ldots \\ x_{2+(i+1)K} &= S^{(2)} \cdot x_{2+iK}, \quad i = 1, 2, \ldots \\ &\vdots \\ x_{K+(i+1)K} &= S^{(K)} \cdot x_{K+iK}, \quad i = 1, 2, \ldots \end{aligned} \tag{4.3}$$

The behaviour of these systems will thus require the eigenvalues and eigenvectors of the *periodic matrix products* $S^{(k)}$. An implicit decomposition of these matrices is now obtained as follows. One proves that there exist unitary matrices Q_i, Z_i, $i = 1, ..., K$ such that :

$$\begin{aligned} \hat{B}_1 &= Z_1^* \cdot B_1 \cdot Q_2 & \hat{A}_1 &= Z_1^* \cdot A_1 \cdot Q_1 \\ \hat{B}_2 &= Z_2^* \cdot B_2 \cdot Q_3 & \hat{A}_2 &= Z_2^* \cdot A_2 \cdot Q_2 \\ &\vdots & & \\ \hat{B}_K &= Z_K^* \cdot B_K \cdot Q_{K+1} & \hat{A}_K &= Z_K^* \cdot A_K \cdot Q_K \end{aligned} \tag{4.4}$$

where now all matrices $\hat{B}_i$, $\hat{A}_i$ are upper triangular and $Q_{K+1} = Q_1$. It is easy to see that in fact the matrices Q_i transform the vectors x_i to $\hat{x}_i = Q_i^* \cdot x_i$ and the matrices $S^{(i)}$ to $\hat{S}^{(i)} = Q_i^* \cdot S^{(i)} \cdot Q_i$. The latter are in upper Schur form, such that the "hat" versions of the systems (8) are now all simultaneously in upper triangular form. Notice that the diagonal elements of the $\hat{S}^{(i)}$ matrices are all equal since they are the products of the diagonal elements of the upper triangular matrices $\hat{B}_i^{-1}\hat{A}_i$. So, if one matrix $\hat{S}^{(i)}$ has a particular ordering of eigenvalues then all other matrices $\hat{S}^{(j)}$ have the same ordering of eigenvalues. In [3] we give an algorithm to compute the above decomposition *implicitly*, i.e. without ever forming the products $S^{(i)}$. The algorithm is based on a preliminary reduction to Hessenberg triangular form where all matrices are put in triangular form except one, which is put in Hessenberg form. We then show the direct connection to the standard QR algorithm, which is sufficient to prove the convergence of the present algorithm. Moreover we show how to reorder the eigenvalues of these Schur forms. We called this the *periodic Schur factorization* [3] because of its relation to periodic systems (6).

Similar unpublished ideas are being pursued by John Hench, UC Santa Barbara (personal communication), who arrives at the same decomposition (11) with a different algorithm. His condensed form essentially consists of all A_i matrices in Hessenberg form and all B_i matrices in triangular form.

We feel that the connection with the QR algorithm then fails to go through, although he reports a good convergence of that algorithm as well.

The application of this decomposition to control theory is apparent. Periodic discrete time systems naturally arise when performing multirate sampling of continuous time systems [10]. In optimal control of such a periodic system one considers the problem :

$$\text{(4.5)} \quad \begin{array}{c} \text{Minimize } J = \sum_{i=1}^{\infty} z_i^T Q_i z_i + u_i^T R_i u_i \\ \text{subject to } E_i z_{i+1} = F_i z_i + G_i u_i \end{array}$$

where the matrices Q_i, R_i, E_i, F_i, G_i are periodic with period K. The Hamiltonian equations are periodic homogenous systems of difference equations (8) in the state z_i and co-state λ_i of the system. The correspondences with (8) are :

$$\text{(4.6)} \quad x_i \doteq \begin{bmatrix} \lambda_i \\ z_i \end{bmatrix}, B_i \doteq \begin{bmatrix} -G_i R_i^{-1} G_i^T & E_i \\ F_i^T & 0 \end{bmatrix}, A_i \doteq \begin{bmatrix} 0 & F_i \\ E_i^T & Q_i \end{bmatrix}.$$

For finding the periodic solutions to the underlying periodic Riccati equation one has to find the stable invariant subspaces of matrices $S^{(i)}$ as above, which happen to be simplectic in the discrete time case (one has to assume here that E_i, F_i and R_i are invertible and eliminate implicitly E_i [13]). Clearly the Schur form is useful here as well as the reordering of eigenvalues [24], [30].

In pole placement of periodic systems [16], again the periodic Schur form and reordering is useful when one wants to extend Varga's pole placement algorithm [37] to periodic systems. Consider the system

$$\text{(4.7)} \quad \begin{array}{c} B_i z_{i+1} = A_i z_i + D_i u_i \\ \text{with state feedback } u_i = F_i z_i + v_i \end{array}$$

where the matrices A_i, B_i, D_i, F_i are periodic with period K. This results in the closed loop system

$$\text{(4.8)} \quad B_i z_{i+1} = (A_i + D_i F_i) z_i + D_i v_i$$

of which the underlying time invariant eigenvalues are those of the matrix :

$$\text{(4.9)} S_F^{(1)} \doteq B_K^{-1}(A_K + D_K F_K) \cdots B_2^{-1}(A_2 + D_2 F_2) B_1^{-1}(A_1 + D_1 F_1).$$

In the above equation it is not apparent at all how to choose the matrices F_i to assign particular eigenvalues of $S_F^{(1)}$. Yet when the matrices A_i, B_i are in the triangular form (11), one can choose the F_i matrices to have only nonzero elements in the last column. This will preserve the triangular form of the matrices $A_i + D_i F_i$ and it is then trivial to choose e.g.

one such column vector to assign one eigenvalue. In order to assign the other eigenvalues one needs to *reorder* the diagonal elements in the periodic Schur form and each time assign another eigenvalue with the same technique. This algorithm will of course fail when the periodic system is not controllable, but this very procedure can in fact be adapted to precisely construct the controllable subspace of the periodic system.

Other applications of this decomposition include periodic Lyapunov equations and robust control of periodic systems. Applications to cyclic Markov chains and two point boundary value problems are also reported in [3]. Another decomposition recently proposed for a sequence of matrices is the generalized QR-decomposition [7], which is connected to the computation of singular values of such sequences. Applications of this in control has not fully been explored yet, but some indications are given in [7]. They inlude the computation of geometric concepts for time varying discrete time systems.

5. Numerical shortcuts and normalized coprime factorizations. A typical approach for solving matrix problems for which there is no "direct" algorithm, is to break it down into a sequence of "intermediate" problems for which one can apply known techniques. We first give a few typical example of this in linear algebra and then extend these ideas to control theory. The particular example of normalized coprime factorizations is then worked out.

An early example of numerical detour was the construction of the normal equations $A^TA \cdot x = A^Tb$ for solving the least squares problem :

$$min||Ax - b||_2. \tag{5.1}$$

This can be "shortcut" by the QR decomposition [15], which is known to be more reliable in general. The sensitivity of the normal equations can only be worse than that of the least squares problem and the numerical stability of the QR decomposition is superior to that of the Choleski decomposition applied to $A^TA \cdot x = A^Tb$. Another typical "detour" is the construction of the matrix $B^{-1}A$ when computing the generalized eigenvalues or generalized singular values of the matrix pair A, B. Nowadays there are implicit decompositions of such matrix pairs which directly yield their generalized eigenvalues and singular values without constructing $B^{-1}A$ [15].

Early examples of numerical shortcuts in control theory include the Schur and generalized Schur methods for Riccati equations [24], [30] and staircase forms for computing various concepts in geometric systems theory [29]. But there are still many other "detours" around that require direct approaches with better numerical properties. We give here an example from robust control. Modern control theory has been largely influenced by the recent development of H_∞ techniques. The underlying theory is now rather well understood but the computational techniques are lagging behind. The design of the robust controller consists in solving factorization problems of transfer matrices and/or related state-space matrix equations [1]. Examples of transfer function factorizations needed here are all-pass factor extraction, inner outer factorization and normalized coprime factorization. Numerical algorithms for such factorizations have been analyzed from a state space and generalized state space point of view [24] [33]. It was found that simple recursive solutions can be obtained for most of them when starting from state space models in so-called condensed forms (such as Schur or generalized Schur forms). The advantage of this approach is a reduced computational complexity and good numerical properties in the corresponding algorithms (see [24], [33] and references therein).

One example where we can expect to shortcut currently proposed algorithms is the construction of normalized coprime factorizations of a given transfer function $R(s) = N(s) \cdot D^{-1}(s)$ with

$$D^*(-\bar{s}) \cdot D(s) + N^*(-\bar{s}) \cdot N(s) = I. \tag{5.2}$$

Presently, one constructs first a stable coprime factorization (with respect to some region Γ), and then one normalizes it via a spectral factorization problem. A more direct approach consists in noticing that $D(s)$ and $N(s)$ must be submatrices of a Γ-stable all-pass $U(s)$ that displays the kernel of $[R(s) \mid -I]$ as follows :

$$\begin{bmatrix} R(s) & -I \end{bmatrix} \cdot U(s) \doteq$$

$$\begin{bmatrix} R(s) & -I \end{bmatrix} \cdot \begin{bmatrix} D(s) & \tilde{N}(s) \\ N(s) & \tilde{D}(s) \end{bmatrix} = \begin{bmatrix} \tilde{R}(s) & 0 \end{bmatrix}. \tag{5.3}$$

This problem now becomes one of rank factorization with all-pass factors and ideas of [33] could be used to tackle this. Notice that this approach avoids "squaring up" the original problem and then performing a spectral factorization. The squared version requires the solution of a Riccati equations where all eigenvalues occur in pairs, whereas this direct decomposition requires the solution of an eigenvalue problem of only half that size. Moreover, just as for the normal equation example, one should expect better numerical sensitivity properties for the direct approach. Similar remarks were already made in [33] for the problems of inner-outer factorizations and other coprime factorizations. The approach proposed there involved eigenvalue problems of typically lower order by using particular state space decompositions.

6. Concluding remarks. The topics discussed in previous sections all point to the significant role linear algebra problems play in systems and control theory. Over the years, numerous algorithms have been developed in that area. Because of the increasing complexity of the problems being tackled, some of them have become challenging from a numerical point of view as well. The interdisciplinary field of numerical linear algebra and linear system theory has lead to some significant developments in the last decade and several of these results are nowadays being implemented in software for CACSD [36], [40]. However, numerical methods in this area are still far from complete. Above we indicated some of these :

- Most of the techniques available today are specifically aimed at dense systems. Only few methods are available that deal with sparse matrices. These are typically based on Krylov type techniques [20], [4], [25]. More work ought to be done in that area using new results as e.g. the QMR method [11].
- There are very few numerically reliable methods for polynomial system models, although typically algorithms in that area are fast. The complexity of these agorithms is usually lower because the underlying matrix problems are highly structured. Analyzing stability of algorithms for structured matrices needs special care as indicated in [34], but use of iterative refinement ought to be useful (see also [19]).
- New decompositions are being found in linear algebra that are particularly relevant to various control problems. Schur forms and generalized Schur forms have already extensively been applied [24], [29], [30]. Also condensed forms have been used for various control problems [31]. But similar extensions for generalized singular values [17], and decompositions involving sequences of matrices have not fully been explored yet [7], [3].
- Many detours are still present in control algorithms because of the absence of direct methods or appropriate decompositions. This is particularly true for problems in robust control, since there one has only recently been able to reduce the problems to known ones involving Riccati equations [14], [8]. Only little attention has been paid yet to the development of appropriate numerical techniques for these specialized problems.

The above list is by far not exhaustive. Other recent developments include e.g. :

- Control algorithms developed for special architectures [5], [13], [6] exploiting parallelism.
- Software developments such as interactive packages [40] and software libraries [36]
- Combinations of the above issues, such as in the problem of model reduction of large sparse systems. This clearly involves sparse matrix techniques but all three other issues come up as well as e.g. indicated in [35].

Acknowledgement

Part of this research was performed while visiting the Institute of Mathematics and Applications of the University of Minnesota, Minneapolis, during the summer quarter of the Applied Linear Algebra Year organized there. We greatly appreciated the hospitality and the productive atmosphere of that institute. We also acknowledge the support of the Research Board of the University of Illinois at Urbana-Champaign (Grant P 1-2-68114) and of the National Science Foundation (Grant CCR 9209349).

REFERENCES

[1] B.D.O. Anderson, Y. Liu, Controller reduction: concepts and approaches, *IEEE Trans. Aut. Contr.* **AC-34** (1989) 802-812.

[2] S. Barnett, *Polynomials and linear control systems*, Marcel Dekker, New York, 1983.

[3] A. Bojanczyk, G. Golub, P. Van Dooren, *The perodic Schur decomposition. Algorithms and applications*, Int. Rept. Stanford Univ., Stanford CA 94305, 1992, also Proceedings SPIE Conference, San Diego, July 1992.

[4] D. Boley, G. Golub, The Lanczos-Arnoldi algorithm and controllability, *Systems and Control Letters* **4** (1984) 317-324.

[5] J.-P. Charlier, P. Van Dooren, A systolic algorithm for Riccati and Lyapunov equations, *Mathematics of Control, Signals and Systems* **2** (1989) 109-136.

[6] B.N. Datta, Parallel algorithms in control theory, *Proceedings 30th IEEE Conf. Dec. & Contr.* (1991) 1700-1704.

[7] B. De Moor, P. Van Dooren, Generalizations of the singular value and QR decomposition, *SIAM Matr. Anal. & Applic.* **13** (1992) to appear.

[8] J. Doyle, B. Francis and A. Tannenbaum, *Feedback Control Theory*, McMillan, 1992.

[9] A. Edelman, *The first annual large dense linear system survey*, Int. Rept. Univ. California, Berkeley CA 94720, 1991.

[10] B. Francis, T. Georgiou, Stability theory for linear time-invariant plants with periodic digital controllers, *IEEE Trans. Aut. Contr.* **33** (1988) 820-832.

[11] R. Freund, N. Nachtigal, QMR: a quasi-minimal residual method for non-Hermitian linear systems, *Numer. Math.* **60** (1991) 315-339.

[12] F.R. Gantmacher, *The Theory of Matrices I and II*, Chelsea, New York, 1959.

[13] J. Gardiner, *Iterative and parallel algorithms for the solution of algebraic Riccati equations*, Ph. D. Thesis, Univ. California, Santa Barbara, 1988.

[14] K. Glover, D. Mc Farlane, Robust stabilization of normalized coprime factor plant descriptions with H_∞-bounded uncertainty, *IEEE Trans. Aut. Contr.* **AC-34** (1989) 821-830.

[15] G. Golub, C. Van Loan, *Matrix Computations* 2nd edition, The Johns Hopkins University Press, Baltimore, Maryland, 1989.

[16] O. Grasselli, S. Longhi, The geometric approach for linear periodic discrete-time systems, *Lin. Algebra & Applic.* **158** (1991) 27-60.

[17] M. Heath, A. Laub, C. Paige, R. Ward, Computing the singular value decomposition of a product of two matrices, *SIAM J. Sci. Stat. Comput.* **7** (1986) 1147-1159.

[18] A. Heemink, Kalman filtering of two-dimensional shallow water flow on a vector processor, *Supercomputer* **20/21** (1987) 32-44.

[19] D. Higham, N. Higham, Backward error and condition of structured linear systems, *SIAM Matrix Anal. & Appl.* **13** (1992) 162-176.

[20] S. Hodel, K. Poolla, Numerical solution of very large sparse Lyapunov equations through approximate power iteration, *Proceedings 29th IEEE Conf. Dec. & Contr.* (1990) 291-296.

[21] M. Jenkins, J. Traub, A three-stage algorithm for real polynomials using quadratic iteration, *SIAM Numer. Anal.* **7** (1970) 545-566.

[22] P. Kokotovic, W. Perkins, J. Cruz, G. D'Ans, ϵ-coupling method for near-optimum design of large-scale linear systems, *Proceedings of the IEE* **5** (1969) 889-892.

[23] V. Kucera, *Discrete linear control. The polynomial equation approach*, Chichester : Wiley & Sons, 1979.

[24] A. Laub, Invariant subspace methods for the numerical solution of Riccati equations, in *The Riccati equation*, Eds. Bittanti, Laub, Willems, Springer Verlag, 1990.

[25] Y. Saad, Overview of Krylov subspace methods with applications to control problems, in *Signal Processing, Scattering and Operator Theory, and Numerical Methods*, Birkhäuser Boston Inc. 1990.

[26] Y. Saad, Analysis of some Krylov subspace approximations to the matrix exponential operator, *SIAM Numer. Anal.* **29** : 209-228 (1992).

[27] R. Srikant, T. Basar, Sequential decomposition and policy iteration schemes for M-player games with partial weak coupling, *Automatica* (1992) to appear.

[28] M. Steinbuch, J. Terlouw, O. Bosgra, S. Smit, Multivariable application of μ analysis of servo system design in the consumer electronics industry, *IEE Colloquium on Successful Industrial Applications of Multivariable Analysis*, London, 1990.

[29] P. Van Dooren, The generalized eigenstructure problem in linear system theory, *IEEE Trans. Aut. Contr.* **AC-26** (1981) 111-129.

[30] P. Van Dooren, A generalized eigenvalue approach for solving Riccati equations, *SIAM Sci. & Stat. Comp.* **2** (1981) 121-135.

[31] P. Van Dooren, M. Verhaegen, On the use of unitary state-space transformations, in *Special Issue of Contemporary Mathematics on Linear Algebra and its Role in Linear System Theory*, AMS, 1985.

[32] P. Van Dooren, Numerical aspects of system and control algorithms, *Journal A* **30** (1989) 25-32.

[33] P. Van Dooren, Rational and polynomial matrix factorizations via recursive pole-zero cancellation, *Linear Algebra & Applications* **137/138** (1990) 663-697.

[34] P. Van Dooren, Structured linear algebra problems in digital signal processing, in *Numerical linear algebra, digital signal processing and parallel algorithms*, Nato Series F, Springer Verlag (1990) 361-384.

[35] P. Van Dooren, Numerical linear algebra techniques for large scale matrix problems in systems and control, *Proceedings 31st IEEE Conf. Dec. & Contr.* (1992) to appear.

[36] A. van den Boom, A. Brown, F. Dumortier, A. Geurts, S. Hammarling, R. Kool, M. Vanbegin, P. Van Dooren, S. Van Huffel, SLICOT, a subroutine library for control and system theory, *IFAC Symp. on CADCS*, Swansea UK, July 1991.

[37] A. Varga, A Schur method for pole assignment, *IEEE Trans. Aut. Contr.* **AC-26** (1981) 517-519.

[38] M. Vidyasagar, *Control System Synthesis : a Factorization Approach*, MIT Press, Cambridge, MA, 1985.

[39] Z. Vostry, New algorithm for polynomial spectral factorization with quadratic convergence I. *Kybernetika* **11** (1975) 415-422.

[40] R. Walker, C. Gregory, S. Shah, MATRIX$_x$: A data analysis, system identification, control design and simulation package, *Control Systems Magazine* **2:4** (1982) 30-37.

[41] J. H. Wilkinson, *The algebraic eigenvalue problem*, Clarendon press, Oxford, 1965.